DESIGN EXTENSION CONDITIONS
AND THE CONCEPT OF
PRACTICAL ELIMINATION
IN THE DESIGN OF
NUCLEAR POWER PLANTS

The following States are Members of the International Atomic Energy Agency:

AFGHANISTAN	GAMBIA	NORWAY
ALBANIA	GEORGIA	OMAN
ALGERIA	GERMANY	PAKISTAN
ANGOLA	GHANA	PALAU
ANTIGUA AND BARBUDA	GREECE	PANAMA
ARGENTINA	GRENADA	PAPUA NEW GUINEA
ARMENIA	GUATEMALA	PARAGUAY
AUSTRALIA	GUINEA	PERU
AUSTRIA	GUYANA	PHILIPPINES
AZERBAIJAN	HAITI	POLAND
BAHAMAS	HOLY SEE	PORTUGAL
BAHRAIN	HONDURAS	QATAR
BANGLADESH	HUNGARY	REPUBLIC OF MOLDOVA
BARBADOS	ICELAND	ROMANIA
BELARUS	INDIA	RUSSIAN FEDERATION
BELGIUM	INDONESIA	RWANDA
BELIZE	IRAN, ISLAMIC REPUBLIC OF	SAINT KITTS AND NEVIS
BENIN	IRAQ	SAINT LUCIA
BOLIVIA, PLURINATIONAL	IRELAND	SAINT VINCENT AND
STATE OF	ISRAEL	THE GRENADINES
BOSNIA AND HERZEGOVINA	ITALY	SAMOA
BOTSWANA	JAMAICA	SAN MARINO
BRAZIL	JAPAN	SAUDI ARABIA
BRUNEI DARUSSALAM	JORDAN	SENEGAL
BULGARIA	KAZAKHSTAN	SERBIA
BURKINA FASO	KENYA	SEYCHELLES
BURUNDI	KOREA, REPUBLIC OF	SIERRA LEONE
CABO VERDE	KUWAIT	SINGAPORE
CAMBODIA	KYRGYZSTAN	SLOVAKIA
CAMEROON	LAO PEOPLE'S DEMOCRATIC	SLOVENIA
CANADA	REPUBLIC	SOUTH AFRICA
CENTRAL AFRICAN	LATVIA	SPAIN
REPUBLIC	LEBANON	SRI LANKA
CHAD	LESOTHO	SUDAN
CHILE	LIBERIA	SWEDEN
CHINA	LIBYA	SWITZERLAND
COLOMBIA	LIECHTENSTEIN	SYRIAN ARAB REPUBLIC
COMOROS	LITHUANIA	TAJIKISTAN
CONGO	LUXEMBOURG	THAILAND
COSTA RICA	MADAGASCAR	TOGO
CÔTE D'IVOIRE	MALAWI	TONGA
CROATIA	MALAYSIA	TRINIDAD AND TOBAGO
CUBA	MALI	TUNISIA
CYPRUS	MALTA	TÜRKİYE
CZECH REPUBLIC	MARSHALL ISLANDS	TURKMENISTAN
DEMOCRATIC REPUBLIC	MAURITANIA	UGANDA
OF THE CONGO	MAURITIUS	UKRAINE
DENMARK	MEXICO	UNITED ARAB EMIRATES
DJIBOUTI	MONACO	UNITED KINGDOM OF
DOMINICA	MONGOLIA	GREAT BRITAIN AND
DOMINICAN REPUBLIC	MONTENEGRO	NORTHERN IRELAND
ECUADOR	MOROCCO	UNITED REPUBLIC OF TANZANIA
EGYPT	MOZAMBIQUE	UNITED STATES OF AMERICA
EL SALVADOR	MYANMAR	URUGUAY
ERITREA	NAMIBIA	UZBEKISTAN
ESTONIA	NEPAL	VANUATU
ESWATINI	NETHERLANDS	VENEZUELA, BOLIVARIAN
ETHIOPIA	NEW ZEALAND	REPUBLIC OF
FIJI	NICARAGUA	VIET NAM
FINLAND	NIGER	YEMEN
FRANCE	NIGERIA	ZAMBIA
GABON	NORTH MACEDONIA	ZIMBABWE

The Agency's Statute was approved on 23 October 1956 by the Conference on the Statute of the IAEA held at United Nations Headquarters, New York; it entered into force on 29 July 1957. The Headquarters of the Agency are situated in Vienna. Its principal objective is "to accelerate and enlarge the contribution of atomic energy to peace, health and prosperity throughout the world".

IAEA SAFETY STANDARDS SERIES No. SSG-88

DESIGN EXTENSION CONDITIONS AND THE CONCEPT OF PRACTICAL ELIMINATION IN THE DESIGN OF NUCLEAR POWER PLANTS

SPECIFIC SAFETY GUIDE

INTERNATIONAL ATOMIC ENERGY AGENCY
VIENNA, 2024

COPYRIGHT NOTICE

© IAEA, 2024

Printed by the IAEA in Austria
January 2024
STI/PUB/2055
https://doi.org/10.61092/iaea.la1m-dy8m

IAEA Library Cataloguing in Publication Data

Names: International Atomic Energy Agency.
Title: Design extension conditions and the concept of practical elimination in the design of nuclear power plants / International Atomic Energy Agency.
Description: Vienna : International Atomic Energy Agency, 2024. | Series: IAEA safety standards series, ISSN 1020-525X ; no. SSG-88 | Includes bibliographical references.
Identifiers: IAEAL 23-01641 | ISBN 978-92-0-130323-3 (paperback : alk. paper) | ISBN 978-92-0-130423-0 (pdf) | ISBN 978-92-0-130523-7 (epub)
Subjects: LCSH: Nuclear power plants — Design and construction. | Nuclear power plants — Safety measures. | Nuclear facilities.
Classification: UDC 621.039.58 | STI/PUB/2055

FOREWORD

by Rafael Mariano Grossi
Director General

The IAEA's Statute authorizes it to "establish…standards of safety for protection of health and minimization of danger to life and property". These are standards that the IAEA must apply to its own operations, and that States can apply through their national regulations.

The IAEA started its safety standards programme in 1958 and there have been many developments since. As Director General, I am committed to ensuring that the IAEA maintains and improves upon this integrated, comprehensive and consistent set of up to date, user friendly and fit for purpose safety standards of high quality. Their proper application in the use of nuclear science and technology should offer a high level of protection for people and the environment across the world and provide the confidence necessary to allow for the ongoing use of nuclear technology for the benefit of all.

Safety is a national responsibility underpinned by a number of international conventions. The IAEA safety standards form a basis for these legal instruments and serve as a global reference to help parties meet their obligations. While safety standards are not legally binding on Member States, they are widely applied. They have become an indispensable reference point and a common denominator for the vast majority of Member States that have adopted these standards for use in national regulations to enhance safety in nuclear power generation, research reactors and fuel cycle facilities as well as in nuclear applications in medicine, industry, agriculture and research.

The IAEA safety standards are based on the practical experience of its Member States and produced through international consensus. The involvement of the members of the Safety Standards Committees, the Nuclear Security Guidance Committee and the Commission on Safety Standards is particularly important, and I am grateful to all those who contribute their knowledge and expertise to this endeavour.

The IAEA also uses these safety standards when it assists Member States through its review missions and advisory services. This helps Member States in the application of the standards and enables valuable experience and insight to be shared. Feedback from these missions and services, and lessons identified from events and experience in the use and application of the safety standards, are taken into account during their periodic revision.

I believe the IAEA safety standards and their application make an invaluable contribution to ensuring a high level of safety in the use of nuclear technology. I encourage all Member States to promote and apply these standards, and to work with the IAEA to uphold their quality now and in the future.

THE IAEA SAFETY STANDARDS

BACKGROUND

Radioactivity is a natural phenomenon and natural sources of radiation are features of the environment. Radiation and radioactive substances have many beneficial applications, ranging from power generation to uses in medicine, industry and agriculture. The radiation risks to workers and the public and to the environment that may arise from these applications have to be assessed and, if necessary, controlled.

Activities such as the medical uses of radiation, the operation of nuclear installations, the production, transport and use of radioactive material, and the management of radioactive waste must therefore be subject to standards of safety.

Regulating safety is a national responsibility. However, radiation risks may transcend national borders, and international cooperation serves to promote and enhance safety globally by exchanging experience and by improving capabilities to control hazards, to prevent accidents, to respond to emergencies and to mitigate any harmful consequences.

States have an obligation of diligence and duty of care, and are expected to fulfil their national and international undertakings and obligations.

International safety standards provide support for States in meeting their obligations under general principles of international law, such as those relating to environmental protection. International safety standards also promote and assure confidence in safety and facilitate international commerce and trade.

A global nuclear safety regime is in place and is being continuously improved. IAEA safety standards, which support the implementation of binding international instruments and national safety infrastructures, are a cornerstone of this global regime. The IAEA safety standards constitute a useful tool for contracting parties to assess their performance under these international conventions.

THE IAEA SAFETY STANDARDS

The status of the IAEA safety standards derives from the IAEA's Statute, which authorizes the IAEA to establish or adopt, in consultation and, where appropriate, in collaboration with the competent organs of the United Nations and with the specialized agencies concerned, standards of safety for protection of health and minimization of danger to life and property, and to provide for their application.

With a view to ensuring the protection of people and the environment from harmful effects of ionizing radiation, the IAEA safety standards establish fundamental safety principles, requirements and measures to control the radiation exposure of people and the release of radioactive material to the environment, to restrict the likelihood of events that might lead to a loss of control over a nuclear reactor core, nuclear chain reaction, radioactive source or any other source of radiation, and to mitigate the consequences of such events if they were to occur. The standards apply to facilities and activities that give rise to radiation risks, including nuclear installations, the use of radiation and radioactive sources, the transport of radioactive material and the management of radioactive waste.

Safety measures and security measures[1] have in common the aim of protecting human life and health and the environment. Safety measures and security measures must be designed and implemented in an integrated manner so that security measures do not compromise safety and safety measures do not compromise security.

The IAEA safety standards reflect an international consensus on what constitutes a high level of safety for protecting people and the environment from harmful effects of ionizing radiation. They are issued in the IAEA Safety Standards Series, which has three categories (see Fig. 1).

Safety Fundamentals

Safety Fundamentals present the fundamental safety objective and principles of protection and safety, and provide the basis for the safety requirements.

Safety Requirements

An integrated and consistent set of Safety Requirements establishes the requirements that must be met to ensure the protection of people and the environment, both now and in the future. The requirements are governed by the objective and principles of the Safety Fundamentals. If the requirements are not met, measures must be taken to reach or restore the required level of safety. The format and style of the requirements facilitate their use for the establishment, in a harmonized manner, of a national regulatory framework. Requirements, including numbered 'overarching' requirements, are expressed as 'shall' statements. Many requirements are not addressed to a specific party, the implication being that the appropriate parties are responsible for fulfilling them.

Safety Guides

Safety Guides provide recommendations and guidance on how to comply with the safety requirements, indicating an international consensus that it

---

[1] See also publications issued in the IAEA Nuclear Security Series.

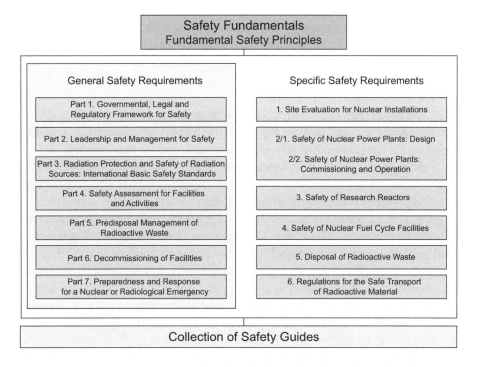

The long term structure of the IAEA Safety Standards Series.

Safety Fundamentals
Fundamental Safety Principles

General Safety Requirements

Part 1. Governmental, Legal and Regulatory Framework for Safety

Part 2. Leadership and Management for Safety

Part 3. Radiation Protection and Safety of Radiation Sources: International Basic Safety Standards

Part 4. Safety Assessment for Facilities and Activities

Part 5. Predisposal Management of Radioactive Waste

Part 6. Decommissioning of Facilities

Part 7. Preparedness and Response for a Nuclear or Radiological Emergency

Specific Safety Requirements

1. Site Evaluation for Nuclear Installations

2/1. Safety of Nuclear Power Plants: Design

2/2. Safety of Nuclear Power Plants: Commissioning and Operation

3. Safety of Research Reactors

4. Safety of Nuclear Fuel Cycle Facilities

5. Disposal of Radioactive Waste

6. Regulations for the Safe Transport of Radioactive Material

Collection of Safety Guides

FIG. 1. The long term structure of the IAEA Safety Standards Series.

is necessary to take the measures recommended (or equivalent alternative measures). The Safety Guides present international good practices, and increasingly they reflect best practices, to help users striving to achieve high levels of safety. The recommendations provided in Safety Guides are expressed as 'should' statements.

APPLICATION OF THE IAEA SAFETY STANDARDS

The principal users of safety standards in IAEA Member States are regulatory bodies and other relevant national authorities. The IAEA safety standards are also used by co-sponsoring organizations and by many organizations that design, construct and operate nuclear facilities, as well as organizations involved in the use of radiation and radioactive sources.

The IAEA safety standards are applicable, as relevant, throughout the entire lifetime of all facilities and activities — existing and new — utilized for peaceful purposes and to protective actions to reduce existing radiation risks. They can be used by States as a reference for their national regulations in respect of facilities and activities.

The IAEA's Statute makes the safety standards binding on the IAEA in relation to its own operations and also on States in relation to IAEA assisted operations.

The IAEA safety standards also form the basis for the IAEA's safety review services, and they are used by the IAEA in support of competence building, including the development of educational curricula and training courses.

International conventions contain requirements similar to those in the IAEA safety standards and make them binding on contracting parties. The IAEA safety standards, supplemented by international conventions, industry standards and detailed national requirements, establish a consistent basis for protecting people and the environment. There will also be some special aspects of safety that need to be assessed at the national level. For example, many of the IAEA safety standards, in particular those addressing aspects of safety in planning or design, are intended to apply primarily to new facilities and activities. The requirements established in the IAEA safety standards might not be fully met at some existing facilities that were built to earlier standards. The way in which IAEA safety standards are to be applied to such facilities is a decision for individual States.

The scientific considerations underlying the IAEA safety standards provide an objective basis for decisions concerning safety; however, decision makers must also make informed judgements and must determine how best to balance the benefits of an action or an activity against the associated radiation risks and any other detrimental impacts to which it gives rise.

DEVELOPMENT PROCESS FOR THE IAEA SAFETY STANDARDS

The preparation and review of the safety standards involves the IAEA Secretariat and five Safety Standards Committees, for emergency preparedness and response (EPReSC) (as of 2016), nuclear safety (NUSSC), radiation safety (RASSC), the safety of radioactive waste (WASSC) and the safe transport of radioactive material (TRANSSC), and a Commission on Safety Standards (CSS) which oversees the IAEA safety standards programme (see Fig. 2).

All IAEA Member States may nominate experts for the Safety Standards Committees and may provide comments on draft standards. The membership of the Commission on Safety Standards is appointed by the Director General and includes senior governmental officials having responsibility for establishing national standards.

A management system has been established for the processes of planning, developing, reviewing, revising and establishing the IAEA safety standards.

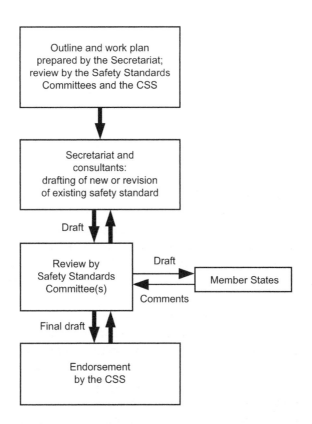

FIG. 2. The process for developing a new safety standard or revising an existing standard.

It articulates the mandate of the IAEA, the vision for the future application of the safety standards, policies and strategies, and corresponding functions and responsibilities.

INTERACTION WITH OTHER INTERNATIONAL ORGANIZATIONS

The findings of the United Nations Scientific Committee on the Effects of Atomic Radiation (UNSCEAR) and the recommendations of international expert bodies, notably the International Commission on Radiological Protection (ICRP), are taken into account in developing the IAEA safety standards. Some safety standards are developed in cooperation with other bodies in the United Nations system or other specialized agencies, including the Food and Agriculture Organization of the United Nations, the United Nations Environment Programme, the International Labour Organization, the OECD Nuclear Energy Agency, the Pan American Health Organization and the World Health Organization.

INTERPRETATION OF THE TEXT

Safety related terms are to be understood as defined in the IAEA Nuclear Safety and Security Glossary (see https://www.iaea.org/resources/publications/iaea-nuclear-safety-and-security-glossary). Otherwise, words are used with the spellings and meanings assigned to them in the latest edition of The Concise Oxford Dictionary. For Safety Guides, the English version of the text is the authoritative version.

The background and context of each standard in the IAEA Safety Standards Series and its objective, scope and structure are explained in Section 1, Introduction, of each publication.

Material for which there is no appropriate place in the body text (e.g. material that is subsidiary to or separate from the body text, is included in support of statements in the body text, or describes methods of calculation, procedures or limits and conditions) may be presented in appendices or annexes.

An appendix, if included, is considered to form an integral part of the safety standard. Material in an appendix has the same status as the body text, and the IAEA assumes authorship of it. Annexes and footnotes to the main text, if included, are used to provide practical examples or additional information or explanation. Annexes and footnotes are not integral parts of the main text. Annex material published by the IAEA is not necessarily issued under its authorship; material under other authorship may be presented in annexes to the safety standards. Extraneous material presented in annexes is excerpted and adapted as necessary to be generally useful.

CONTENTS

1. INTRODUCTION

BACKGROUND

1.1. The publication of IAEA Safety Standards Series No. SSR-2/1, Safety of Nuclear Power Plants: Design, in 2012[1] and its subsequent revision in 2016 as SSR-2/1 (Rev. 1) [1], introduced changes to the requirements for the design of nuclear power plants. These changes include measures for strengthening the application of the concept of defence in depth as follows:

(a) Including design extension conditions among the plant states to be considered in the design;
(b) Ensuring by design that plant event sequences that could lead to an early radioactive release or a large radioactive release[2] have been considered for 'practical elimination'[3];
(c) Including design features that enable the use of non-permanent equipment for power supply and cooling.

1.2. The incorporation of these aspects into designs of new nuclear power plants will affect the necessary safety assessment. IAEA Safety Standards Series No. GSR Part 4 (Rev. 1), Safety Assessment for Facilities and Activities [3], establishes requirements for performing the safety assessment for all types of facility and activity, including assessment of defence in depth. Specific requirements for the safety assessment and the safety analysis of nuclear power plants are established in SSR-2/1 (Rev. 1) [1].

OBJECTIVE

1.3. The objective of this Safety Guide is to provide recommendations for the design of new nuclear power plants on the application of selected requirements

---

[1] INTERNATIONAL ATOMIC ENERGY AGENCY, Safety of Nuclear Power Plants: Design, IAEA Safety Standards Series No. SSR-2/1, IAEA, Vienna (2012).

[2] An 'early radioactive release' in this context is a release of radioactive material for which off-site protective actions are necessary but are unlikely to be fully effective in due time. A 'large radioactive release' is a release of radioactive material for which off-site protective actions that are limited in terms of times and areas of application are insufficient for protecting people and the environment [1, 2].

[3] For a definition of the term 'practical elimination', see the Definition section at the end of this publication.

1

in SSR-2/1 (Rev. 1) [1] related to defence in depth and the practical elimination of plant event sequences that could lead to an early radioactive release or a large radioactive release. This Safety Guide also provides recommendations in relation to design aspects of defence in depth, in particular on those aspects associated with design extension conditions.

1.4. This Safety Guide is intended for use by organizations involved in the verification, review and assessment of safety of nuclear power plants. It is also intended to be of use to organizations involved in the design, manufacture, construction, modification and operation of nuclear power plants and in the provision of technical support for nuclear power plants, as well as to regulatory bodies.

SCOPE

1.5. This Safety Guide applies primarily to new land based stationary nuclear power plants with water cooled reactors designed for electricity generation or for other heat production applications (such as district heating or desalination) (see para. 1.6 of SSR-2/1 (Rev. 1) [1]). It is recognized that for reactors cooled by other media or for reactors based on innovative design concepts, some of the recommendations in this Safety Guide might not be applicable or fully applicable, or judgement might be needed in their application.

1.6. For nuclear power plants designed in accordance with earlier standards (see para. 1.3 of SSR-2/1 (Rev. 1) [1]), this Safety Guide might be useful when evaluating potential safety enhancements of such designs, for example as part of the periodic safety review of the plant.

1.7. This Safety Guide focuses on the implementation and assessment of the design safety measures provided in para. 1.1. These measures play an important role in the application of the concept of defence in depth, which constitutes the primary means of both preventing and mitigating the consequences of accidents, in accordance with Principle 8 of IAEA Safety Standards Series No. SF-1, Fundamental Safety Principles [4].

1.8. As described in para. 2.13 of SSR-2/1 (Rev. 1) [1], defence in depth at nuclear power plants comprises five levels. Plant states considered in the design correspond to one or more levels of defence in depth. This Safety Guide is structured in terms of the design of safety provisions[4] necessary for each plant state, rather than for each level of defence in depth. In this way, the significance and the importance of design extension conditions for the safety approach are emphasized. The specific focus of this Safety Guide is on the nuclear fuel, as the main source of radioactivity, with special emphasis on design extension conditions.

1.9. This Safety Guide considers the assessment of the independence of structures, systems and components implemented at different defence in depth levels in a general manner. However, factors that could cause dependence between structures, systems and components, such as environmental factors, operational or human factors, and external or internal hazards, are not addressed in detail in this Safety Guide.

1.10. This Safety Guide does not provide specific recommendations for the design of particular safety features for design extension conditions or for any other plant state considered in the design. Such recommendations are provided in Safety Guides for the design of various types of plant system, such as IAEA Safety Standards Series Nos SSG-56, Design of the Reactor Coolant System and Associated Systems for Nuclear Power Plants [5]; SSG-53, Design of the Reactor Containment and Associated Systems for Nuclear Power Plants [6]; SSG-34, Design of Electrical Power Systems for Nuclear Power Plants [7]; and SSG-39, Design of Instrumentation and Control Systems for Nuclear Power Plants [8].

1.11. This Safety Guide does not consider the specific safety analyses to be carried out for different plant states, as these are addressed in IAEA Safety Standards Series Nos SSG-2 (Rev. 1), Deterministic Safety Analysis for Nuclear Power Plants [9]; SSG-3 (Rev. 1), Development and Application of Level 1 Probabilistic Safety Assessment for Nuclear Power Plants [10]; and SSG-4, Development and Application of Level 2 Probabilistic Safety Assessment for Nuclear Power Plants [11], as appropriate. However, this Safety Guide takes into account the recommendations provided in these publications.

STRUCTURE

1.12. Section 2 sets out the requirements in SSR-2/1 (Rev. 1) [1] that govern the approach to the design of nuclear power plants relating to the prevention of radiological consequences, on which the recommendations in this Safety

---

[4] In this Safety Guide, 'safety provisions' are used to refer to design solutions applied to structures, systems and components and related operational strategies.

Guide are based. Section 3 provides recommendations on the implementation and assessment of design extension conditions within the concept of defence in depth, and on the independence of the safety provisions considered for the levels of defence in depth. Section 4 provides recommendations on the application of the concept of practical elimination of plant event sequences that could lead to an early radioactive release or a large radioactive release. Section 5 provides recommendations on the implementation of design provisions for enabling the use of non-permanent equipment for power supply and cooling.

1.13. Annex I provides examples of cases of practical elimination that may differ among Member States. Annex II provides some considerations for the application of recommendations included in this Safety Guide to nuclear power plants designed to earlier standards (see para. 1.3 of SSR-2/1 (Rev. 1) [1]).

2. DESIGN APPROACH CONSIDERING THE RADIOLOGICAL CONSEQUENCES OF ACCIDENTS

2.1. This Safety Guide focuses on the design features of a nuclear power plant that provide for the protection of the public and the environment in accident conditions and that should be assessed for compliance with a number of requirements in SSR-2/1 (Rev. 1) [1]. These requirements pertain to the general plant design and particularly to the capability of the plant to withstand, without unacceptable radiological consequences, accidents that are either more severe than design basis accidents or that involve additional failures.

2.2. Requirement 5 of SSR-2/1 (Rev. 1) [1] states:

"The design of a nuclear power plant shall be such as to ensure that radiation doses to workers at the plant and to members of the public do not exceed the dose limits, that they are kept as low as reasonably achievable in operational states for the entire lifetime of the plant, and that they remain below acceptable limits and as low as reasonably achievable in, and following, accident conditions."

2.3. Paragraph 4.3 of SSR-2/1 (Rev. 1) [1] states (footnote omitted):

"The design shall be such as to ensure that plant states that could lead to high radiation doses or to a large radioactive release have been 'practically eliminated', and that there would be no, or only minor, potential radiological consequences for plant states with a significant likelihood of occurrence."

2.4. Furthermore, para. 4.4 of SSR-2/1 (Rev. 1) [1] states that (footnote omitted) "Acceptable limits for purposes of radiation protection associated with the relevant categories of plant states shall be established, consistent with the regulatory requirements."

2.5. Further requirements on criteria and objectives relating to radiological consequences of different plant states considered in the design, including accident conditions, are also established in SSR-2/1 (Rev. 1) [1] as follows:

— "Criteria shall be assigned to each plant state, such that frequently occurring plant states shall have no, or only minor, radiological consequences and plant states that could give rise to serious consequences shall have a very low frequency of occurrence" (para. 5.2 of SSR-2/1 (Rev. 1) [1]).
— "A primary objective shall be to manage all design basis accidents so that they have no, or only minor, radiological consequences, on or off the site, and do not necessitate any off-site protective actions" (para. 5.25 of SSR-2/1 (Rev. 1) [1] in relation to design basis accidents).
— "The design shall be such that the possibility of conditions arising that could lead to an early radioactive release or a large radioactive release is 'practically eliminated'" (para. 5.31 of SSR-2/1 (Rev. 1) [1] in relation to design extension conditions).
— "The design shall be such that for design extension conditions, protective actions that are limited in terms of lengths of time and areas of application shall be sufficient for the protection of the public, and sufficient time shall be available to take such measures" (para. 5.31A of SSR-2/1 (Rev. 1) [1] in relation to design extension conditions).

2.6. Paragraph 2.10 of SSR-2/1 (Rev. 1) [1] states:

"Measures are required to be taken to ensure that the radiological consequences of an accident would be mitigated. Such measures include the provision of safety features and safety systems, the establishment of accident management procedures by the operating organization and, possibly, the establishment of off-site protective actions by the appropriate authorities, supported as necessary by the operating organization, to mitigate exposures if an accident occurs."[5]

2.7.   As stated in para. 2.13 of SSR-2/1 (Rev. 1) [1], "The safety objective in the case of a severe accident is that only protective actions that are limited in terms of lengths of time and areas of application would be necessary and that off-site contamination would be avoided or minimized."

2.8.   Harmful radiological consequences to the public can arise only from the occurrence of uncontrolled accidents. Therefore, the recommendations in this Safety Guide are focused on the implementation and assessment of the concept of defence in depth and the complementary need to demonstrate the practical elimination of plant event sequences that could lead to an early radioactive release or a large radioactive release.

2.9.   Recommendations on radiation protection in the design of nuclear power plants are provided in IAEA Safety Standards Series No. SSG-90, Radiation Protection Aspects of Design for Nuclear Power Plants [13]. Recommendations on the protection of the public and the environment are provided in IAEA Safety Standards Series No. GSG-8, Radiation Protection of the Public and the Environment [14].

---

[5]   The establishment of off-site protective actions belongs to the fifth level of defence in depth and is outside the scope of this Safety Guide. Requirements regarding such arrangements are established in IAEA Safety Standards Series No. GSR Part 7, Preparedness and Response for a Nuclear or Radiological Emergency [12].

3. IMPLEMENTATION AND ASSESSMENT OF DESIGN EXTENSION CONDITIONS WITHIN THE CONCEPT OF DEFENCE IN DEPTH

OVERALL IMPLEMENTATION OF DEFENCE IN DEPTH

3.1. The concept of defence in depth for the design of nuclear power plants is described in paras 2.12–2.14 of SSR-2/1 (Rev. 1) [1]. As stated in para. 2.14 of SSR-2/1(Rev. 1) [1]:

"A relevant aspect of the implementation of defence in depth for a nuclear power plant is the provision in the design of a series of physical barriers, as well as a combination of active, passive and inherent safety features that contribute to the effectiveness of the physical barriers in confining radioactive material at specified locations. The number of barriers that will be necessary will depend upon the initial source term in terms of the amount and isotopic composition of radionuclides, the effectiveness of the individual barriers, the possible internal and external hazards, and the potential consequences of failures."

3.2. Requirement 7 of SSR-2/1 (Rev. 1) [1] on the application of defence in depth in the design of nuclear power plants states that **"The design of a nuclear power plant shall incorporate defence in depth. The levels of defence in depth shall be independent as far as is practicable."** Paragraphs 4.9–4.13A of SSR-2/1 (Rev. 1) [1] develop this overarching requirement.

3.3. For the safety provisions at each level of defence in depth, the following should be demonstrated:

(a) The performance of the safety provisions implemented at that level to maintain the integrity of the barriers;
(b) The adequate reliability of the safety provisions at that level so that it can be assured, with a sufficient level of confidence, that a certain plant condition can be brought under control without the need to implement safety provisions associated with the next level of defence in depth;

(c)   The independence, as far as practicable, of the safety provisions at that level, including their physical separation[6], from the safety provisions associated with the previous levels of defence in depth.

3.4.   Frequently, for purposes of design safety and operational safety, the various levels of defence in depth are associated with the various plant states considered in the design. The introduction of design extension conditions among the plant states has resulted in different approaches in different States regarding interpretation of the correspondence between the plant states considered in the design and the levels of defence in depth. Two of these approaches are presented in Table 1.

3.5.   In Approach 1, depicted on the left side of Table 1, design extension conditions without significant fuel degradation are associated with the third level of defence in depth. With this approach, each level has a clear objective that reflects the progression of an accident and the protection of the barriers (i.e. the third level is implemented to prevent fuel damage and the fourth level is implemented to mitigate severe accidents and prevent off-site contamination). As stated in para. 3.39 of SSG-2 (Rev. 1) [9]:

"The initial selection of sequences for design extension conditions without significant fuel degradation should be based on the consideration of single initiating events of very low frequency or multiple failures to meet the acceptance criteria with regard to the prevention of core damage."

TABLE 1. LEVELS OF DEFENCE IN DEPTH

Level of defence	Objective	Essential design means	Essential operational means	Level of defence
Approach 1				Approach 2
Level 1	Prevention of abnormal operation and failures	Robust design and high quality in construction of normal operation systems, including monitoring and control systems	Operational limits and conditions and normal operating procedures	Level 1

---

[6] Physical separation is separation by geometry (distance, orientation, etc.), by appropriate barriers, or by a combination thereof [2].

8

TABLE 1. LEVELS OF DEFENCE IN DEPTH (cont.)

Level of defence Approach 1		Objective	Essential design means	Essential operational means	Level of defence Approach 2
Level 2		Control of abnormal operation and detection of failures	Limitation and protection systems and other surveillance features	Abnormal operating procedures and/or emergency operating procedures	Level 2
Level 3	3a	Control of design basis accidents	Safety systems	Emergency operating procedures	Level 3
	3b	Control of design extension conditions to prevent core melting	Safety features for design extension conditions without significant fuel degradation[a]	Emergency operating procedures	Level 4
Level 4		Control of design extension conditions to mitigate the consequences of severe accidents	Safety features for design extension conditions with core melting[b] Technical support centre	Severe accident management guidelines	
Level 5		Mitigation of the radiological consequences of significant releases of radioactive substances	On-site and off-site emergency response facilities	On-site and off-site emergency plans and procedures	Level 5

[a] Such safety features are understood as additional safety features for design extension conditions, or as safety systems with an extended capability to prevent the consequences of severe accidents (see para. 5.27 of SSR-2/1 (Rev. 1) [1]).

[b] Such safety features are understood as additional safety features for design extension conditions, or as safety systems with an extended capability to mitigate the consequences of severe accidents or to maintain the integrity of the containment (see para. 5.27 of SSR-2/1 (Rev. 1) [1]).

Therefore, in Approach 1, acceptable limits on predicted radiological consequences for design extension conditions without significant fuel degradation may be the same as, or similar to, acceptable limits for design basis accidents. Furthermore, the physical phenomena associated with design basis accidents and design extension conditions without significant fuel degradation are similar, although there might be differences in the analysis. In contrast, the physical phenomena associated with design extension conditions with core melt are completely different.

3.6. In Approach 2, depicted on the right side of Table 1, design extension conditions without significant fuel degradation and design extension conditions with core melt are considered together in the fourth level of defence in depth. This approach emphasizes the distinction between the set of rules to be applied for design extension conditions and the set of rules to be applied for design basis accidents, both in the design and in the safety assessment.

3.7. Despite their differences, both approaches are in compliance with para. 5.29(a) of SSR-2/1 (Rev. 1) [1] and support the implementation, to the extent practicable, of independence between safety systems and those safety features for design extension conditions.

Normal operation and anticipated operational occurrences

3.8. Operational states comprise two sets of plant states: normal operation and anticipated operational occurrences. Modes of normal operation include, for example, startup, power operation, shutting down, shutdown and refuelling and are defined in the documentation governing the operation of the plant (e.g. the operational limits and conditions[7]). Anticipated operational occurrences[8] could happen from a postulated initiating event[9] involving a failure to prevent an abnormal operation or an equipment failure expected to happen during the operating lifetime of the plant.

---

[7] In some States, the term 'technical specifications' is used instead of the term 'operational limits and conditions'.

[8] An anticipated operational occurrence is a deviation of an operational process from normal operation that is expected to occur at least once during the operating lifetime of a facility but which, in view of appropriate design provisions, does not cause any significant damage to items important to safety or lead to accident conditions [2].

[9] Examples of relevant postulated initiating events are provided in para. 3.28 of SSG-2 (Rev. 1) [9].

3.9. Paragraph 4.13 of SSR-2/1 (Rev. 1) [1] states:

"The design shall be such as to ensure, as far as is practicable, that the first, or at most the second, level of defence is capable of preventing an escalation to accident conditions for all failures or deviations from normal operation that are likely to occur over the operating lifetime of the nuclear power plant."

Therefore, to maintain the integrity of the first physical barrier for the confinement of radioactive substances (i.e. the fuel cladding) and to prevent a significant release of primary coolant, design provisions for operational states should have adequate capabilities to achieve the following:

(a) To prevent failures or deviations from normal operation by means of robust design, compliance with proven engineering practices and high quality standards commensurate with the importance to safety of these design provisions;

(b) To detect and intercept deviations from normal operation and return the plant to a state of normal operation;

(c) To prevent anticipated operational occurrences, once they start, from escalating into accident conditions.

3.10. The reliability of safety provisions for anticipated operational occurrences should be such that the frequency of transition to a design basis accident is lower than the highest frequency of postulated initiating events for design basis accidents (usually lower than 10^{-2} per reactor-year) (see table II–1 in annex II to SSG-2 (Rev. 1) [9]).

Design basis accidents

3.11. Requirement 19 of SSR-2/1 (Rev. 1) [1] states:

"A set of accidents that are to be considered in the design shall be derived from postulated initiating events for the purpose of establishing the boundary conditions for the nuclear power plant to withstand, without acceptable limits for radiation protection being exceeded."

3.12. Paragraph 5.24 of SSR-2/1 (Rev. 1) [1] states that "Design basis accidents shall be used to define the design bases, including performance criteria, for safety systems and for other items important to safety that are necessary to control design basis accident conditions."

3.13. Paragraph 5.25 of SSR-2/1 (Rev. 1) [1] states:

"The design shall be such that for design basis accident conditions, key plant parameters do not exceed the specified design limits. A primary objective shall be to manage all design basis accidents so that they have no, or only minor, radiological consequences, on or off the site, and do not necessitate any off-site protective actions."

Consequently, specific design provisions (i.e. safety systems) should be implemented to prevent and mitigate the radiological consequences of design basis accidents by preventing significant fuel damage and maintaining the integrity of the containment (i.e. by preserving the structural integrity of the containment and maintaining its associated systems[10]). The objective of the safety systems is to limit the radiological consequences for the public and the environment to the extent that no off-site protective actions are necessary.

3.14. The accident conditions that are most likely to occur during the lifetime of a plant are categorized as design basis accidents and should have an expected frequency typically below 10^{-2} per reactor-year. Design basis accidents should include single initiating events[11] due to failure of the first and the second levels of defence in depth. The safety systems should be designed to control postulated initiating events considered for design basis accidents by ensuring that safety functions can be fulfilled and barriers can be maintained. The safety systems designed to control design basis accidents requiring prompt and reliable action (see para. 5.11 of SSR-2/1 (Rev. 1) [1]) should rely on automatic actuation, and the need for short term operator actions should be minimized. The safety systems should be designed, constructed and maintained to ensure reliability commensurate with their safety significance. Safety design concepts, such as adequate margins and redundancy, are required to be applied in the design and construction of the safety systems (see Requirement 24 and paras 5.21A, 5.42 and 5.73 of SSR-2/1 (Rev. 1) [1]). The environmental conditions considered in the qualification programme of the safety systems should correspond to the loads and adverse environmental conditions induced by design basis accidents and postulated internal and external hazards. Further recommendations on the design of specific safety systems for nuclear power plants are provided in the corresponding Safety Guides (see SSG-56 [5], SSG-53 [6], SSG-34 [7] and SSG-39 [8]).

---

[10] The containment and its associated systems are described in para. 1.3 of SSG-53 [6].

[11] In some States, the term 'infrequent and limiting faults' is used (see table II-1 in annex II to SSG-2 (Rev. 1) [9]), while other States may use different terms.

Design extension conditions

3.15. Requirement 20 of SSR-2/1 (Rev. 1) [1] states:

"**A set of design extension conditions shall be derived on the basis of engineering judgement, deterministic assessments and probabilistic assessments for the purpose of further improving the safety of the nuclear power plant by enhancing the plant's capabilities to withstand, without unacceptable radiological consequences, accidents that are either more severe than design basis accidents or that involve additional failures. These design extension conditions shall be used to identify the additional accident scenarios to be addressed in the design and to plan practicable provisions for the prevention of such accidents or mitigation of their consequences.**"

3.16. Paragraph 5.30 of SSR-2/1 (Rev. 1) [1] states:

"In particular, the containment and its safety features shall be able to withstand extreme scenarios that include, among other things, melting of the reactor core. These scenarios shall be selected by using engineering judgement and input from probabilistic safety assessments."

3.17. To meet the requirements presented in paras 3.15 and 3.16, two separate categories of design extension conditions[12] can be identified: design extension conditions without significant fuel degradation[13] and design extension conditions with core melting.[14] This distinction reflects the fact that the most frequent design extension conditions should not lead to a fuel degradation, in accordance with the objective of prevention of fuel degradation.

3.18. As presented in Table 1 and in paras 3.4–3.7, the following two main approaches for design extension conditions are used by States:

(a)   In some States, design extension conditions are divided into design extension conditions without significant fuel degradation and design extension

---

[12] The definition of 'design extension conditions' is provided in SSR-2/1 (Rev. 1) [1].

[13] The term 'design extension conditions without significant fuel degradation' comprises situations to be analysed for the fuel in the reactor core and for the fuel in the spent fuel pool.

[14] In some States, these categories of design extension conditions are denoted respectively as 'design extension conditions A' (without significant fuel degradation) and 'design extension conditions B' (with core melting).

conditions with core melting. In some States, very low frequency initiating events are treated as design extension conditions without significant fuel degradation. In other States, design extension conditions without significant fuel degradation are postulated for complex sequences involving multiple failures, whereas very low frequency postulated single initiating events are treated as design basis accidents. Recommendations related to design extension conditions without significant fuel degradation are provided in paras 3.19–3.28. Recommendations related to design extension conditions with core melting are provided in paras 3.29–3.36.

(b) In some States, design extension conditions are not subdivided on the basis of fuel condition or number of failures. In this approach, the same high level dose limits and analysis rules are used for all event sequences of design extension conditions. States using this approach may use the recommendations provided in paras 3.19–3.36 as appropriate.

Design extension conditions without significant fuel degradation

3.19. A process for the comprehensive identification of design extension conditions without significant fuel degradation should be developed. Paragraphs 3.39–3.44 of SSG-2 (Rev. 1) [9] provide recommendations for the identification of design extension conditions without significant fuel degradation.

3.20. In general, the control of design extension conditions without significant fuel degradation should be accomplished by safety features specifically designed and qualified for such conditions. Alternatively, design extension conditions without significant fuel degradation can be controlled by available safety systems, provided that the safety systems have not been affected by the events that led to the design extension conditions under consideration and that they are capable and qualified to operate under the associated environmental conditions. Requirement 13 of SSR-2/1 (Rev. 1) [1] states that **"Plant states shall be identified and shall be grouped into a limited number of categories primarily on the basis of their frequency of occurrence at the nuclear power plant."**

3.21. The deterministic safety analyses of design basis accidents and design extension conditions without significant fuel degradation may share similar safety objectives, namely to demonstrate that the integrity of barriers will be maintained and to prevent core damage or damage to the fuel in the spent fuel pool (see paras 7.28 and 7.45 of SSG-2 (Rev. 1) [9]).

3.22. Design basis accidents and design extension conditions without significant fuel degradation are also distinguished in terms of the application of different design requirements and the use of different acceptable limits or criteria[15] or approaches for performing deterministic safety analysis. Thus, for design extension conditions without significant fuel degradation, the following apply:

(a) Less stringent design requirements than for design basis accidents might be applied. For example, safety features for design extension conditions without significant fuel degradation may be assigned to a lower safety class than safety systems.

(b) Less conservative assumptions than for design basis accidents, or best estimate methods, are acceptable for the deterministic safety analysis (see paras 7.35–7.44 and 7.47–7.55 of SSG-2 (Rev. 1) [9]).

(c) The requirements for the overall limits or criteria related to the radiological consequences for design extension conditions are established in para. 5.31A of SSR-2/1 (Rev. 1) [1]. States may choose to apply more restrictive limits or criteria for design extension conditions without significant fuel degradation. For example, some States may choose to apply identical or similar overall limits or criteria for radiological consequences to those for design basis accidents (see paras 7.32, 7.33 and 7.46 of SSG-2 (Rev. 1) [9]).

3.23. If it is possible to use available safety systems to respond to design extension conditions without significant fuel degradation, safety analysis is still required to demonstrate their effectiveness (see Requirement 42 of SSR-2/1 (Rev. 1) [1]). The deterministic safety analysis may use less conservative methods and assumptions than for design basis accidents (see para. 3.22). Nevertheless, there should still be an adequate level of confidence in the results of the deterministic safety analysis, and the safety margins to avoid cliff edge effects should be demonstrated to be adequate (see paras 7.45, 7.54 and 7.55 of SSG-2 (Rev. 1) [9]).

3.24. Design basis accidents are required to be analysed in a conservative manner (see para. 5.26 of SSR-2/1 (Rev. 1) [1]). However, design extension conditions without significant fuel degradation have the potential to exceed the established capabilities of safety systems. Therefore, it might be possible to show that some safety systems, with an extended capability in their design, would be capable of, and be qualified for, mitigating the design extension conditions without significant fuel degradation, on the basis of best estimate analyses and less conservative assumptions than the assumptions used for design basis accidents.

---

[15] 'Acceptable limits related to radiological consequences' used in SSR-2/1 (Rev. 1) [1] and 'acceptance criteria related to radiological consequences' used in SSG-2 (Rev. 1) [9] are considered to be equivalent terms.

3.25. Similarly to design basis accidents, radioactive releases should be minimized as far as reasonably achievable for design extension conditions without significant fuel degradation.

3.26. Anticipated operational occurrences and frequent design basis accidents combined with failures in safety systems should be considered as part of the list of design extension conditions without significant fuel degradation (see para. 3.40 of SSG-2 (Rev. 1) [9]). In many plant designs, such conditions include anticipated transient without scram and station blackout[16].

3.27. On the basis of engineering judgement and of deterministic safety analyses and probabilistic safety assessments, design extension conditions without significant fuel degradation should also be considered in the identification of safety provisions to be implemented to prevent and reduce the frequency of severe accidents caused by failures of safety systems. Such safety provisions should include, if possible, additional, diverse measures to cope with common cause failures of safety systems.

3.28. Consideration of design extension conditions without significant fuel degradation reinforces the robustness of the design to cope with some complex and unlikely failure sequences and balances the overall risk profile of the plant. Therefore, the reliability of safety systems and of safety features for design extension conditions without significant fuel degradation should be sufficiently high to prevent a severe accident by making the escalation to a severe accident very unlikely to occur.

Design extension conditions with core melting

3.29. In accordance with Requirement 42 and paras 5.9 and 5.30 of SSR-2/1 (Rev. 1) [1], and with consideration of results from research and development, a set of representative accident conditions with core melting should be postulated to provide inputs for the design of the containment and of the safety features ensuring its functionality. This set of representative accident conditions should be considered in the design of safety features for design extension conditions with core melting and should represent bounding cases that envelop other severe accidents with more limited degradation of the core.

---

[16] See para. 5.8 of SSG-34 [7] for the definition of station blackout.

3.30. Paragraph 6.68 of SSR-2/1 (Rev. 1) [1] states (footnote omitted):

"For reactors using a water pool system for fuel storage, the design shall be such as to prevent the uncovering of fuel assemblies in all plant states that are of relevance for the spent fuel pool so that the possibility of conditions arising that could lead to an early radioactive release or a large radioactive release is 'practically eliminated' and so as to avoid high radiation fields on the site."

Hence, significant fuel degradation in the spent fuel pool should not be postulated as part of this set of design extension conditions; rather it is required to be considered among the conditions to be practically eliminated (see Section 4).

3.31. A detailed analysis should be performed and documented to identify and characterize accident conditions that could lead to core damage and also challenge or bypass the containment. Relevant accident conditions that could lead to core damage should be postulated as design extension conditions (see paras 3.46 and 3.47 of SSG-2 (Rev. 1) [9] and para. 2.11 of SSG-53 [6]), even though the design provisions taken in accordance with the requirements of SSR-2/1 (Rev. 1) [1] to prevent such accidents will make the probability of core damage very low. Aspects that affect the accident progression and that influence the containment response and the source term should be taken into account in the design of safety features for design extension conditions with core melting (see para. 3.42 of SSG-53 [6]).

3.32. The capability and the reliability of the safety features for design extension conditions with core melting should be evaluated to ensure that they are adequate for the safety function that they need to fulfil.

3.33. The challenges to plant safety presented by design extension conditions with core melting, and the extent to which the design may be reasonably expected to mitigate their consequences, should be considered in establishing procedures and guidelines for accident management. Recommendations in this regard are provided in IAEA Safety Standards Series No. SSG-54, Accident Management Programmes for Nuclear Power Plants [15].

3.34. In order to avoid the risk to the integrity of the containment resulting from overpressurization, the pressure inside the containment should be controlled. This may be achieved by ensuring and maintaining adequate cooling of the reactor containment atmosphere during design extension conditions with core melting, by a filtered reactor containment venting system allowing the containment pressure to be reduced, or by other design features or alternative measures (see para. 11.8

of SSG-90 [13]). The consequences of filtered and unfiltered direct leakage of radioactive releases from the reactor containment in design extension conditions with core melting should remain below the design target defined in accordance with the recommendations provided in para. 2.7 of SSG-53 [6] and para. 2.10 of SSG-90 [13], and assessed in accordance with the recommendations provided in para. 11.7 of SSG-90 [13] to allow sufficient time for the implementation of off-site protective actions. At any time, radioactive releases should be controlled to meet the timing and magnitude criteria for avoiding radioactive releases considered as an early radioactive release or a large radioactive release.

3.35. As stated in paras 3.44 and 3.45 of SSG-53 [6]:

"Multiple means to control the pressure buildup in accident conditions inside the containment should be implemented, and venting (if any [is included in the design]) should be used only as a last resort. ... [T]he use of the venting system should not lead to an early radioactive release or a large radioactive release".

3.36. A safety assessment of the design should be performed with consideration of the progression of severe accident phenomena and their consequences and the achievement of acceptable end state conditions, and should take into account applicable topical issues. More detailed information on the range of physical processes that could occur following core damage is provided in para. 7.66 of SSG-2 (Rev. 1) [9].

ASSESSMENT OF THE IMPLEMENTATION OF DEFENCE IN DEPTH

3.37. An assessment of the implementation of defence in depth in the design of a nuclear power plant is required in order to ensure that the safety provisions for each level are adequately designed to meet the objectives of that level in terms of prevention, detection, limitation and mitigation. Requirement 13 of GSR Part 4 (Rev. 1) [3] states that **"It shall be determined in the assessment of defence in depth whether adequate provisions have been made at each of the levels of defence in depth."**

3.38. Paragraphs 4.45–4.48A of GSR Part 4 (Rev. 1) [3] establish additional requirements for the assessment of defence in depth.

3.39. The performance and reliability of safety provisions for all plant states should be assessed, taking into consideration an applicable set of analysis rules, the level of risk and the safety significance of the safety provisions. The safety provisions should be designed to maintain the integrity of the barriers to the extent necessary for the relevant plant state, or to mitigate the consequences of postulated failures. The assessment should provide evidence that the performance and reliability of the safety provisions associated with each level of defence in depth are adequate. The assessment should demonstrate that, for each credible initiating event, the risk is commensurate with the frequency of the event, also considering all consequences of internal hazards and external hazards that could cause the event. The assessment should consider insights from the assessment of engineering aspects and from deterministic safety analysis and probabilistic safety assessment, as appropriate for each different plant state.

3.40. The multiplicity of the levels of defence is not a justification to weaken the effectiveness of some levels by relying on the effectiveness of other levels. In a sound and balanced design, structures, systems and components at each level of defence are characterized by a reliability commensurate with their function and their safety significance, and reasonable safety margins are provided.

3.41. The defence in depth concept should be applied for all sources of radiation present in the nuclear power plant. The following are examples of sources of radiation likely to be present in a nuclear power plant:

— The reactor core;
— Fresh nuclear fuel, irradiated fuel and fuel casks;
— Neutron sources and other radioactive sources;
— Airborne radioactive substances in buildings;
— Piping and process equipment containing radioactive substances (e.g. the reactor coolant system; reactor cooling systems; auxiliary systems; heating, ventilation and air conditioning systems of controlled areas; gas and liquid effluent treatment systems; solid waste treatment systems).

3.42. For sources of radiation other than the reactor core and the nuclear fuel, defence in depth should be implemented in accordance with a graded approach, with account taken of the fact that some levels of defence in depth might not be appropriate for many sources of radiation within the plant. Account should be taken of the risk represented by the amount and type of radioactive substance present; the potential for its dispersion owing to its physical and chemical nature; and the possibility of nuclear, chemical or thermal reactions that could occur under normal or abnormal conditions and the kinetics of such reactions. These characteristics will differ for different sources of radiation and will influence the necessary number of levels of defence in depth and the strength of each level.

3.43. The physical barriers included in the design are an important consideration when assessing the adequacy of the implementation of defence in depth. For each identified source of radiation, the physical barriers should be identified and their robustness should be evaluated in accordance with a graded approach. The following aspects should be assessed in the evaluation:

(a) Each barrier should be designed with an appropriate margin, and the robustness of the various barriers should be evaluated by applying a graded approach based on the radiation risks or the safety class of the equipment forming the barrier.
(b) Appropriate codes and standards should be used for the design and manufacture or construction of barriers, and proven materials and technologies should be used in the manufacture or construction.
(c) All loads and combinations of loads that can apply to the barriers in operational states and accident conditions, including loads caused by the effects of the internal hazards and external hazards considered in the design, should be identified and calculated and should be shown to be less than the applicable limits.
(d) The number of barriers provided in the design should be justified and the barriers chosen for each plant state should offer the best protection for workers and the public that may be reasonably expected.
(e) Valves, their control equipment and other equipment used in the barriers to prevent radioactive releases should be designed to ensure structural integrity of the barriers in accident conditions.
(f) Any deviation of a barrier from its normal configuration (e.g. open containment to accommodate certain activities when the plant is in a shutdown state) should be justified by demonstrating that adequate protection is maintained despite the temporary configuration (or operation) of the barrier.

3.44. An analysis of the various mechanisms that could challenge or degrade the performance of the safety functions should be carried out in order to assess the adequacy of the safety provisions that are implemented to prevent the occurrence of such mechanisms or to stop their progression. To the extent that different degradation mechanisms could necessitate different safety provisions, the adequacy and effectiveness of each safety provision should be assessed for each degradation mechanism.

3.45. The adequacy and effectiveness of safety provisions should be assessed by performing deterministic safety analyses that model the plant response to a given initiating event for different boundary conditions representative of each plant state. Each plant state should be characterized by a type of deterministic safety analysis, with an applicable set of analysis rules, level of conservatism and acceptance criteria. Recommendations on conducting deterministic safety analyses for the different plant states are provided in SSG-2 (Rev. 1) [9].

3.46. The performance of safety provisions at each level of defence in depth is assessed through the assessment of engineering aspects and by deterministic safety analysis involving the use of validated and verified computer codes and models to demonstrate that acceptance criteria are met and that there are sufficient margins to avoid cliff edge effects. Further recommendations are provided in paras 5.14–5.39 of SSG-2 (Rev. 1) [9].

3.47. The reliability analysis of safety provisions for the different plant states, as indicated in para. 3.39, typically uses probabilistic techniques and takes into account the plant layout and either protective provisions against or qualification for the effects of hazards, as well as potential commonalities in the design, manufacture, maintenance and testing of redundant and diverse equipment.

3.48. Statements of reliability should be supported by equipment reliability data that are shown to be relevant to the structure, system or component being assessed, as well as supported by test data, the use of proven technologies and engineering practices, and feedback from operating experience. Statements of reliability should also be supported by verification of compliance of the structure, system or component with the applicable set of design requirements. Reliability analyses for different systems or levels of defence in depth can be integrated into a probabilistic safety assessment to evaluate overall plant risk metrics, such as core damage frequency or frequencies of early radioactive releases or large radioactive releases.

3.49. It should be verified that adequate diversity has been implemented in the design of systems fulfilling the same fundamental safety function in different plant states if a common cause failure of those systems would result in unacceptable damage to the fuel or unacceptable radiological consequences.

3.50. The reliability of structures, systems and components for controlling anticipated operational occurrences should be such that they effectively reduce challenges to safety systems and contribute to preventing the occurrence of accident conditions.

3.51. The reliability of both safety systems and additional safety features for design extension conditions without significant fuel degradation should be such that the core damage frequency does not exceed any safety goal of the plant, where set (e.g. for new nuclear power plants, typically below 10^{-5} per reactor-year). Design extension conditions without significant fuel degradation should be postulated (see paras 3.39–3.44 of SSG-2 (Rev. 1) [9]) and analysed considering applicable analysis rules (see paras 7.45–7.55 of SSG-2 (Rev. 1) [9]) as appropriate to achieve the safety goals.

3.52. Any vulnerabilities that could result in the complete failure of a safety system should be identified and it should be assessed whether such a failure, in combination with a postulated initiating event, could escalate to a core melt accident. For each such combination analysed, if the consequences exceed those acceptable for design basis accidents and might involve a core melt with unacceptable frequency, safety features that are separate, independent and diverse and unlikely to fail by the same common cause should be implemented (e.g. an alternate AC power supply in case of a total loss of the emergency power supply, or a separate and diverse decay heat removal chain).

3.53. The capability and reliability of safety features for design extension conditions with core melting should be sufficient to ensure that the integrity of the containment will not be jeopardized during any postulated core melt sequence. Any large uncertainties associated with the analyses of core melt accidents should be taken into account when evaluating the reliability of the safety features.

3.54. It should be demonstrated that the reliability of safety systems and safety features for design extension conditions has taken into account the reliability of their supporting systems.

INDEPENDENCE BETWEEN LEVELS OF DEFENCE IN DEPTH

3.55. Paragraph 4.13A of SSR-2/1 (Rev. 1) [1] states:

"The levels of defence in depth shall be independent as far as practicable to avoid the failure of one level reducing the effectiveness of other levels. In particular, safety features for design extension conditions (especially features for mitigating the consequences of accidents involving the melting of fuel) shall as far as is practicable be independent of safety systems."

3.56. Some additional requirements in SSR-2/1 (Rev. 1) [1] contribute to ensuring the independence of the levels of defence in depth. For example, the sharing of structures, systems or components for executing functions in different plant states is one factor that could compromise the independence of the levels of defence in depth. Requirement 21 of SSR-2/1 (Rev. 1) [1] states:

"**Interference between safety systems or between redundant elements of a system shall be prevented by means such as physical separation, electrical isolation, functional independence and independence of communication (data transfer), as appropriate.**"

3.57. For protection systems and control systems, in particular, Requirement 64 of SSR-2/1 (Rev. 1) states that "**Interference between protection systems and control systems at the nuclear power plant shall be prevented by means of separation, by avoiding interconnections or by suitable functional independence.**"

3.58. Regarding supporting systems and auxiliary systems, Requirement 69 of SSR-2/1 (Rev. 1) [1] states that "**The design of supporting systems and auxiliary systems shall be such as to ensure that the performance of these systems is consistent with the safety significance of the system or component that they serve at the nuclear power plant.**"

3.59. The potential for common cause failures is a second factor that can compromise the independence of the levels of defence in depth. Typical root causes of common cause failures are undetected human errors in design or manufacturing, human errors in the operation or maintenance, inadequate equipment qualification or inadequate protection against internal or external hazards. Requirement 24 of SSR-2/1 (Rev. 1) [1] states:

"**The design of equipment shall take due account of the potential for common cause failures of items important to safety, to determine how the concepts of diversity, redundancy, physical separation and functional independence have to be applied to achieve the necessary reliability.**"

3.60. Full independence of the levels of defence in depth may be difficult to achieve. The design of a nuclear power plant should consider all potential causes of dependencies and an approach should be implemented to remove them to the extent reasonably practicable. Robust independence should be implemented among systems whose simultaneous failure would result in conditions that have harmful effects on people or the environment.

3.61. As far as practicable, the sharing of safety systems or parts of them for executing safety functions for different plant states should be avoided. In particular, it should be ensured that within the event sequence that might follow a postulated initiating event, a safety system credited to respond in a given plant state will not have been needed for a preceding plant state. As stated in para. 4.13A of SSR/2-1 (Rev. 1) [1], "safety features for design extension conditions (especially features for mitigating the consequences of accidents involving the melting of fuel) shall as far as is practicable be independent of safety systems."

3.62. The systems needed for different plant states should be functionally isolated from one another in such a way that a malfunction or failure in a system in a given plant state does not affect another system needed in a different plant state. However, practical limitations of the reactor design may in certain situations necessitate exemptions to such functional isolation, although each case should be justified.

3.63. The systems intended for mitigating design extension conditions with core melting should be functionally and physically separated from the systems intended for other plant states to the extent practicable. However, safety features for design extension conditions with core melting may, for good reasons, also be used for preventing severe core damage, if it can be demonstrated that such use will not undermine the ability of these safety features to perform their primary function if conditions do evolve into design extension conditions with core melting. As an example, a power supply intended for design extension conditions with core melting could be used, if necessary, to power equipment for design extension conditions without significant fuel degradation.

ASSESSMENT OF THE INDEPENDENCE OF THE LEVELS OF DEFENCE IN DEPTH

3.64. Engineering assessment and deterministic and probabilistic methods should be used to assess the independence of the levels of defence in depth. The structures, systems and components needed for each postulated initiating event should be identified, and it should be shown by means of engineering analyses that the structures, systems and components needed for implementing each level of defence in depth are sufficiently independent from those for the other levels. A postulated initiating event is generally a bounding event covering different kinds of initiating failure and so it might be difficult to list all equipment for normal operation that might initially be affected by the postulated initiating event for particular design extension conditions. For this reason, the crediting of systems for normal operation in the safety assessment of design extension conditions should be considered with extreme caution and should be adequately justified.

The adequacy of the independence between levels of defence in depth should also be assessed by probabilistic analyses.

3.65. The assessment should demonstrate that independence between successive levels of defence is adequate to limit the progression of deviations from normal operation and to prevent harmful effects on the public and the environment if an accident occurs. The assessment of the independence of the levels of defence in depth should aim to verify that the vulnerabilities for common cause failures between structures, systems and components claimed to be independent have been identified and removed to the extent practicable. Such common cause failures might have originated in the layout, design, manufacture, operation or maintenance. If a functional dependency between structures, systems and components has not been removed, this should be justified in the assessment.

3.66. The assessment should demonstrate that safety systems that are intended to respond in an accident are not jeopardized by the initiating event. The assessment should demonstrate that the operability of the safety systems is not jeopardized by failures in systems designed for normal operation. Following an initiating event, the failures occurring in anticipated operational occurrences should not compromise the capability of the safety systems to manage a design basis accident.

3.67. The assessment should demonstrate that a failure of a supporting system is not capable of simultaneously affecting parts of systems for different plant states in a way that the capability to fulfil a safety function is compromised. For this purpose, the assessment should provide evidence that the reliability, redundancy, diversity and independence of supporting systems are commensurate with the significance to safety of the system being supported.

3.68. An assessment should be conducted of the independence of structures, systems and components that might be necessary at different levels of defence in depth to mitigate the consequences of a single hazard or a likely combination of internal or external hazards on the plant. It should be demonstrated that the postulated initiating event and the failures induced in the plant cannot result in common cause failure of the structures, systems and components necessary for mitigation of the consequences of the hazard at different levels of defence in depth. In particular, the assessment should be conducted to ensure that a common cause failure will not affect at the same time (i) the safety functions performed by the safety systems or some safety features for design extension conditions without significant fuel degradation and (ii) the safety functions of the necessary safety features for design extension conditions with core melting.

4. PRACTICAL ELIMINATION OF PLANT EVENT SEQUENCES THAT COULD LEAD TO AN EARLY RADIOACTIVE RELEASE OR A LARGE RADIOACTIVE RELEASE

4.1. Paragraph 2.11 of SSR-2/1 (Rev. 1) [1] states (footnote omitted):

"Plant event sequences that could result in high radiation doses or in a large radioactive release have to be 'practically eliminated'… An essential objective is that the necessity for off-site protective actions to mitigate radiological consequences be limited or even eliminated in technical terms, although such measures might still be required by the responsible authorities."

4.2. In relation to the fourth level of defence in depth, para. 2.13 of SSR-2/1 (Rev. 1) [1] states that (footnotes omitted) "Event sequences that would lead to an early radioactive release or a large radioactive release are required to be 'practically eliminated'."

4.3. Paragraph 5.31 of SSR-2/1 (Rev. 1) [1] states that (footnote omitted) "The design shall be such that the possibility of conditions arising that could lead to an early radioactive release or a large radioactive release is 'practically eliminated'."

4.4. Although the term 'early radioactive release' is predominantly used in SSR-2/1 (Rev. 1) [1], the term 'high radiation doses' appears in paras 2.11 and 4.3 of SSR-2/1 (Rev. 1) [1]. It should be interpreted to mean such doses as would occur as a result of an early radioactive release, because protective actions could not be effectively implemented in time to prevent them.

4.5. The concept of practical elimination should be applied only to those events or sequences of events that could lead to unacceptable consequences (i.e. early radioactive release or large radioactive release) that cannot be mitigated by reasonably practicable means. The practical elimination of such plant event sequences is required to be ensured by design (see SSR-2/1 (Rev. 1) [1]), either by ensuring that the plant event sequence is physically impossible (see paras 4.33 and 4.34) or because the plant event sequence is considered, with a high level of confidence, to be extremely unlikely to arise (see paras 4.35–4.42).

4.6. The concept of practical elimination should be applied as part of the overall safety approach to the design of nuclear power plants, as set out in section 2 of

SSR-2/1 (Rev. 1) [1]. As a result of the adequate implementation of the first, second, third and fourth levels of defence in depth, the likelihood of an off-site radioactive release that could potentially result from an accident will be very low for most cases. However, it is necessary to verify that there would not be credible plant conditions that could not be effectively and practicably mitigated and that could thus lead to unacceptable radiological consequences. This is the aim of the practical elimination concept: to complement the adequate implementation of defence in depth at a plant with a focused analysis of those conditions having the potential for unacceptable radiological consequences.

4.7. Practical elimination should not be seen as an alternative to mitigation of the consequences of a severe accident (i.e. implementation of the fourth and fifth levels of defence in depth). Rather, the application of practical elimination may lead (i) to the identification of additional provisions that will complement defence in depth in the design by explicitly identifying those core melt sequences that cannot be reasonably managed, and (ii) to the implementation of additional means to prevent those core melt sequences. Moreover, the practical elimination of plant event sequences that could lead to an early radioactive release or a large radioactive release does not remove the need for emergency preparedness and response in accordance with Principle 9 of SF-1 [4] and the requirements of GSR Part 7 [12].

4.8. SSR-2/1 (Rev. 1) [1] does not provide quantitative acceptance limits or criteria for the radiological consequences of accident conditions, nor for the magnitude of what is to be considered an early radioactive release or a large radioactive release. Independent of the design or of specific definitions of the terms, early radioactive releases or large radioactive releases are those that will challenge provisions of the fifth level of defence in depth. In some States, an early radioactive release is defined for a specific site, considering restrictions on implementing off-site protective actions in a timely manner. In other States, large releases are considered to be releases much larger than core melt acceptance criteria, leading to a very significant impact on the public or the environment. In other States, acceptable limits on radioactive releases for the purpose of radiation protection, and probabilistic criteria or target values for the purpose of demonstrating a low frequency of a core damage accident, have been established, consistent with regulatory requirements or objectives.

4.9. The concept of practical elimination should be applied in a new nuclear power plant from an early stage, when it is more practicable to design and implement additional[17] safety features. The incorporation of such features should be an iterative process, which should use insights from engineering experience and from deterministic safety analyses and probabilistic safety assessments in a complementary manner. Additionally, it is recognized that operational measures may be needed throughout the lifetime of the plant to ensure that the design assumptions are met.

IDENTIFICATION OF RELEVANT PLANT EVENT SEQUENCES

4.10. The first step in demonstrating the practical elimination of plant event sequences that could lead to an early radioactive release or a large radioactive release is the identification of such plant event sequences. This identification process is expected to result in a list of plant event sequences, which can be grouped into a smaller set of plant conditions among the severe accidents identified for the plant. The identification process should be justified and supported by relevant information.

4.11. In a severe accident, large quantities of radioactive substances are present and not confined in the fuel or within the reactor coolant system. In addition, severe accident phenomena can generate large amounts of energy very rapidly. Together, these challenge the confinement of radioactive substances, which might give rise to unacceptable radiological consequences.

4.12. Therefore, if a severe accident occurs, it is necessary to ensure that radioactive substances released from the nuclear fuel will be confined. In particular, in situations of limited confinement (e.g. in accidents involving fuel storage or when the containment is open and cannot be closed in time, or where there is a containment bypass that cannot be isolated), the only way to prevent unacceptable radiological consequences is to prevent the occurrence of such severe accidents. In such cases, it is necessary to demonstrate practical elimination by proving the physical impossibility of the accident or by proving with a high level of confidence that such severe accidents would be extremely unlikely. Therefore, the issue when considering whether a particular plant event sequence should be practically eliminated is the potential for the event sequence to lead to a failure of the confinement function.

---

[17] Such additional safety features include any design provision that is implemented following an assessment supporting the demonstration of practical elimination of some plant event sequences. Some design provisions will already have been implemented to support other safety objectives and analyses and can also support the demonstration of practical elimination.

4.13. To help ensure that the demonstration of practical elimination is manageable, the whole set of individual plant event sequences that might lead to unacceptable radiological consequences should be grouped to form a limited number of bounding cases or types of accident condition (see also para. 4.15). The following five general types of plant event sequence should be considered, depending on their applicability for specific designs:

(a) Plant event sequences that could lead to prompt reactor core damage and consequent early containment failure, such as the following:
(i) Failure of a large pressure retaining component in the reactor coolant system;
(ii) Uncontrolled reactivity accidents.
(b) Plant event sequences that could lead to early containment failure, such as the following:
(i) Highly energetic direct containment heating;
(ii) Large steam explosion;
(iii) Explosion of combustible gases, including hydrogen and carbon monoxide.
(c) Plant event sequences that could lead to late containment failure, such as the following:
(i) Base mat penetration or other damage to the integrity of the containment during molten corium–concrete interaction;
(ii) Long term loss of containment heat removal (e.g. failure of the containment heat removal system);
(iii) Explosion of combustible gases, including hydrogen and carbon monoxide.
(d) Plant event sequences with containment bypass, such as the following:
(i) A loss of coolant accident with the potential to drive the leakage outside of the containment via supporting systems (i.e. a loss of coolant accident in an interface system)[18];
(ii) Plant event sequences producing a consequential containment bypass (e.g. an induced steam generator tube rupture);
(iii) Plant event sequences with core melt, which include spent fuel pool sequences for plants that have a spent fuel pool located inside the

---

[18] As the containment function might be jeopardized by the initiating event, any escalation to significant fuel degradation has to be analysed and, where relevant, considered for practical elimination.

containment, and in which the containment is open[19] (e.g. in the shutdown state).

(e) Significant fuel degradation in a spent fuel pool.

4.14. The grouping in para. 4.13 is consistent with the recommendations provided in para. 3.67 of SSG-53 [6] and para. 3.56 of SSG-2 (Rev. 1) [9] and highlights some examples of plant event sequences for consideration for practical elimination.

4.15. Other criteria for grouping are also possible. The consequences of the accidents in para. 4.13(c)(i) and (ii) could in fact be mitigated by the implementation of reasonable technical means. Also, some bypass sequences in para. 4.13(d) may involve adequate natural retention of radioactive substances to achieve the safety goal. In such cases, for scenarios not retained within the scope of consideration for practical elimination, evidence of the effectiveness and an appropriate reliability of the mitigation should be provided. To facilitate the grouping proposed, each type of plant event sequence should be analysed to identify the associated combination of failures or associated physical phenomena that are specific to the plant design and that have the potential to lead to a loss of the confinement function.

4.16. The identification and grouping described in paras 4.13 and 4.15 should combine, when relevant, the following approaches:

(a) A phenomenological (top-down) approach, in which phenomena are considered that might challenge the confinement function before or in the course of a severe accident, in order to define a comprehensive list of plant event sequences (i.e. as listed in para. 4.13).

(b) A sequence oriented (bottom-up) approach, in which all plant event sequences that could lead to a severe accident are reviewed. For each sequence, any challenge to the confinement function is assessed (which might involve evaluation of the loads on the containment and of possible release routes via leakages and bypasses). The sequence oriented approach supplements the phenomenological approach with broader screening to identify all relevant plant event sequences.

4.17. All possible normal operating modes of the plant (e.g. startup, power operation, shutting down, shutdown and refuelling), including operating modes

---

[19] In many light water reactor designs, the technology used for equipment hatches might not be fast enough to ensure reclosure and restoration of the integrity of the containment before a radioactive release occurs.

with an open containment, should be considered in the process of identifying relevant event sequences.

4.18. All plant locations and buildings where nuclear fuel is stored (including the spent fuel pool) should be considered in the process of identifying relevant plant event sequences.

IDENTIFICATION AND ASSESSMENT OF SAFETY PROVISIONS FOR DEMONSTRATING PRACTICAL ELIMINATION

4.19. The assessment aimed at identifying safety provisions in the form of design and operational features that could be implemented for demonstrating the practical elimination of each relevant plant event sequence should consider the following aspects:

(a) The state of the art in nuclear science and technology, as appropriate;
(b) Experience from the operation of nuclear power plants and from accidents;
(c) Proven technical and industrial feasibility of safety provisions;
(d) The capability of safety provisions to provide sufficient margins for dealing with uncertainties and to avoid cliff edge effects;
(e) Potential drawbacks of safety provisions, which might only become evident after the plant is put into operation (e.g. operational constraints or spurious actuations);
(f) The kinetics of the severe accident phenomena that might pose a risk to the integrity of the containment or its leaktightness;
(g) Means to reduce the need to conduct on-site actions or use off-site personnel or equipment.

4.20. The identification of safety provisions necessitates a comprehensive analysis of the physical phenomena involved from the deterministic, probabilistic and engineering judgement perspectives, and it might be necessary to further refine the identification of event sequences performed in accordance with the approaches described in para. 4.16.

4.21. The designer should establish a decision making process for determining reasonably practicable safety provisions to achieve practical elimination. When several options for safety provisions have been considered, the rationale for selecting the final design of safety provisions should be documented.

4.22. The safety provisions identified to demonstrate the practical elimination of relevant plant event sequences should be associated, on a case by case basis, with the appropriate levels of defence in depth or plant states, in particular those levels at which the event sequence would need to be interrupted to prevent unacceptable radiological consequences. It should be verified that the appropriate engineering design rules (e.g. fail-safe actuation and protection against common cause failures induced by internal and external hazards) and the technical requirements for the safety provisions in that level of defence in depth or plant state have been followed. The aim of this verification is to ensure that the safety provisions would achieve their safety function with sufficient margins to account for uncertainties under the prevailing conditions (e.g. the harsh environmental conditions associated with a severe accident). In applying the engineering design rules and the technical requirements, where relevant, appropriate testing should be applied, operational procedures should be followed, and, in operation, surveillance as well as in-service testing and inspection should be conducted. The engineering design rules and the technical requirements should be applied at all steps in the development of the safety provisions, from design to operation and including their manufacture, construction or implementation at the plant and their commissioning and periodic testing.

4.23. Safety provisions for demonstrating the practical elimination of some severe accident conditions could include the need for design provisions as well as operational provisions, and as such they could involve operator actions (e.g. the opening of primary circuit depressurization valves to prevent high pressure core melt conditions). The number of essential operator actions should be kept low and, when unavoidable, a human factor assessment should be part of the justification supporting any claim for high reliability of operator actions. The human factor assessment should address the following:

(a) The availability of information given to operating personnel to perform the actions from the control room or locally, the quality of the procedures or guidelines to implement the actions, and the training of the operating personnel.

(b) The environment for performing the actions (e.g. access to the local area, components to be handled, identification of the location of components, ambient conditions). If local actions are expected to be taken in harsh environmental conditions, this is likely to reduce the reliability of the demonstration of practical elimination.

(c) The timescales for performing the actions, including sufficient margins to achieve the expected outcomes.

4.24. Some safety provisions claimed to contribute towards the practical elimination of some plant event sequences could be vulnerable to human errors that might have occurred prior to the onset of the accident. Such human errors could introduce latent risks that might prevent successful operation of a system or component when it is called upon during an event or accident. In such cases, the system or component used to perform the action should be subject to relevant operational provisions (e.g. periodic testing, in-service inspection and surveillance, qualification tests following maintenance and periodic system alignment checks) to limit the risk from human errors of this type.

4.25. Paragraph 5.21A of SSR-2/1 (Rev. 1) [1] states:

"The design of the plant shall also provide for an adequate margin to protect items ultimately necessary to prevent an early radioactive release or a large radioactive release in the event of levels of natural hazards exceeding those considered for design, derived from the hazard evaluation for the site."

Therefore, certain safety provisions for demonstrating practical elimination should be designed to withstand relevant internal and external hazards (i.e. hazards that are consequential to the accident condition or likely to arise concurrently) with an appropriate margin.

4.26. Where safety provisions for demonstrating practical elimination rely on support functions, the relevant supporting systems should all be designed to the standards necessary to ensure that they have the same level of reliability as the safety provisions. The design should use a combination of safety design principles such as redundancy, separation, diversity and robustness to hazards to achieve the intended reliability of the relevant safety function. Alternatively, the safety provisions should be tolerant to the loss of support functions.

DEMONSTRATION OF PRACTICAL ELIMINATION

4.27. The overall effectiveness of the safety provisions identified and included to demonstrate practical elimination should be demonstrated through a safety assessment that includes engineering judgement, deterministic analyses and probabilistic assessments. The demonstration of practical elimination should be conducted as part of the design and safety assessment process for the plant, including the necessary inspection and surveillance processes during manufacture, construction, commissioning and operation.

4.28. All safety provisions developed to prevent the occurrence of the plant event sequences in each of the groups in para. 4.13 should be analysed. None of the phenomena or accident conditions indicated should be overlooked because of their low likelihood of occurrence. Credible research results should be used to support claims of effectiveness of the safety provisions.

4.29. For each group of plant event sequences considered for practical elimination, an assessment should be performed to demonstrate the effectiveness of the associated safety provisions. Either it should be demonstrated that it is physically impossible for the event sequence to arise (see paras 4.33 and 4.34) or it should be demonstrated, with a high level of confidence, that the event sequence is extremely unlikely to arise (see paras 4.35–4.42). The justification for the practical elimination of an event sequence should preferably rely on a demonstration of the physical impossibility of its occurrence. If this is not achievable, it should be demonstrated, with a high level of confidence, that such a plant event sequence is extremely unlikely to occur.

4.30. As is evident from para. 4.13, the various plant event sequences to be considered for practical elimination are inherently very different. As a consequence, their practical elimination should be demonstrated on a case by case basis.

4.31. Uncertainties due to limited knowledge of some physical phenomena, in particular severe accident phenomena, should be considered when conducting engineering analyses as well as deterministic safety analyses and probabilistic safety assessments, so that a high level of confidence in the result can be assured.

4.32. Computer codes and calculations used to support the demonstration of practical elimination should be verified and validated, and models used should reflect best understanding of the physical phenomena involved so as to provide an acceptable prediction of the plant event sequences and the phenomena involved. Section 5 of SSG-2 (Rev. 1) [9] provides recommendations on the use of computer codes for deterministic safety analyses.

Practical elimination of plant event sequences because they would be physically impossible

4.33. Where a claim is made that a plant event sequence can be practically eliminated because it is physically impossible, it should be demonstrated that the inherent safety characteristics of the system or reactor type are such that the plant event sequence cannot, by the laws of nature, occur and that the fundamental safety functions (see Requirement 4 of SSR-2/1 (Rev. 1) [1]) will always be fulfilled.

4.34. In practice, the demonstration of physical impossibility is limited to very specific cases (see Annex I). The demonstration of physical impossibility cannot rely on measures that involve active components or operator actions.

Practical elimination of plant event sequences considered, with a high level of confidence, to be extremely unlikely to arise

4.35. The demonstration that certain plant sequences are extremely unlikely to occur should rely on the assessment of engineering aspects and deterministic considerations, supported by probabilistic considerations to the extent practicable, taking into account the uncertainties due to the limited knowledge of some physical phenomena. Although probabilistic targets can be set (e.g. frequencies of core damage or of radioactive releases), the demonstration of practical elimination cannot be approached only by probabilistic means. Probabilistic insights should be used in support of deterministic and engineering analyses. Meeting a probabilistic target alone is not a justification to exclude further deterministic and engineering analyses and possible implementation of additional reasonably practicable safety provisions to reduce the risk. Thus, the low probability of occurrence of an accident with core damage is not a reason for discounting further consideration of means to protect the containment against the conditions generated by such an accident. In contrast, design extension conditions with core melting are required to be postulated in the design, in accordance with para. 5.30 of SSR-2/1 (Rev. 1) [1].

4.36. The demonstration that a plant event sequence can be practically eliminated should consider the following, as applicable:

(a) An adequate set of safety provisions, including both equipment and organizational provisions;
(b) The robustness of these safety provisions (e.g. adequate margins, adequate reliability, qualification for the operational conditions);
(c) The independence between the equipment safety provisions described in points (a) and (b) (i.e. an adequate combination of redundancy, physical separation, diversity and functional independence).

4.37. Deterministic safety analyses of severe accidents should be performed using a realistic approach (see Option 4 in table 1 of SSG-2 (Rev. 1) [9]), to the extent practicable. Because explicit quantification of uncertainties might be impractical owing to the complexity of the phenomena and insufficient experimental data, sensitivity analyses should be performed to demonstrate the robustness of the results and to support the conclusions of the safety analyses. Sensitivity analyses could also be used to confirm the adequacy and representativeness of the selected severe accidents considered for the bounding analysis.

4.38. When probabilistic arguments are used to support a claim that a particular plant event sequence has been practically eliminated, it should be ensured that the cumulative contribution of all the different event sequences considered does not exceed the target frequency for early radioactive releases or large radioactive releases, if such a target has been claimed by the designer or operating organization in the safety assessment of the plant or has been established by the regulatory body.

4.39. The validity of any probabilistic models used should be confirmed for the intended application. Assumptions made in support of this should be well justified and validated.

4.40. The limitations of uncertainties associated with the models used in the demonstration of practical elimination should be identified, taking into account that limitations of probabilistic safety assessments are associated with the probabilistic modelling, as well as the supporting deterministic conservative or best estimate analyses.

4.41. If the plant event sequence to be practically eliminated is the result of a single initiating event, such as the failure of a large pressure retaining component in normal operation, the demonstration of practical elimination should rely on the substantiation that a high level of quality is achieved at all stages of the lifetime of the component (i.e. its design, manufacture, implementation, commissioning and operation, including periodic testing and in-service surveillance, if any) so as to prevent the occurrence and propagation of any defect liable to cause the failure of the component.[20] Hence, both the occurrence of the single initiating event (e.g. the failure of a large pressure retaining component) and the consequential events (i.e. the prompt reactor core damage and consequent early containment failure) should be considered for practical elimination.

4.42. If the plant event sequence to be practically eliminated is the result of an event sequence in which the confinement function is degraded to such an extent that adequate retention of the radioactive substance is not possible before core melt occurs, then it should be demonstrated, with a high level of confidence, that core melt will be prevented. This means that, at a minimum, the usual levels

---

[20] In some States, this demonstration is associated with other concepts such as 'incredibility of failure', 'break preclusion', 'high integrity component' and 'non-breakable component', rather than with the concept of practical elimination.

of defence in depth should be implemented (i.e. for anticipated operational occurrences, design basis accidents and design extension conditions without significant fuel degradation) with enhancements, as necessary, to prevent design extension conditions with core melt.

DOCUMENTATION OF THE APPROACH TO PRACTICAL ELIMINATION

4.43. The safety analysis report of the plant should reflect the measures taken to demonstrate the practical elimination of plant event sequences that could lead to an early radioactive release or a large radioactive release. The safety analysis report should include, either directly or by reference, all elements of the demonstration, including the approach used to identify such event sequences, the design and operational safety provisions implemented to ensure that the possibility of such event sequences arising has been practically eliminated, and the corresponding analyses.

5. IMPLEMENTATION OF DESIGN PROVISIONS FOR ENABLING THE USE OF NON-PERMANENT EQUIPMENT FOR POWER SUPPLY AND COOLING

5.1. As an application of Requirement 14 of SSR-2/1 (Rev. 1) [1], the design basis for items important to safety should take into account the most limiting conditions under which they need to operate or maintain their integrity. This includes the conditions resulting from internal and external hazards. In accordance with Requirement 17 of SSR-2/1 (Rev. 1) [1], the effects of internal and external hazards and relevant combinations of hazards are required to be evaluated. For external hazards, this is done as part of the site evaluation for the plant (see IAEA Safety Standards Series No. SSR-1, Site Evaluation for Nuclear Installations [16]).

5.2. There have been cases in which some natural external hazards, such as extreme earthquakes and tsunamis, have exceeded the levels of external hazards considered for design, derived from the hazard evaluation for the site. Paragraph 5.21A of SSR-2/1 (Rev. 1) [1] states that adequate margins are required to be provided in the design to protect against external hazards for such cases.

5.3. To provide additional resilience against event sequences exceeding those considered as the basis for the design, such as levels of external hazards exceeding those considered for design, derived from the hazard evaluation for the site, several requirements are established in SSR-2/1 (Rev. 1) [1] regarding the inclusion of features in the design to enable the safe use of non-permanent equipment for the following purposes[21]:

(a) Restoring the necessary electrical power supplies (para. 6.45A of SSR-2/1 (Rev. 1) [1]);
(b) Restoring the capability to remove heat from the containment (para. 6.28B of SSR-2/1 (Rev. 1) [1]);
(c) Ensuring sufficient water inventory for the long term cooling of spent fuel and for providing shielding against radiation (para. 6.68 of SSR-2/1 (Rev. 1) [1]).

5.4. The use of non-permanent equipment for other similar purposes, such as the removal of residual heat from the core, is not explicitly required but is not excluded.

5.5. Non-permanent equipment is primarily intended for preventing unacceptable radioactive consequences in the long term phase of accident conditions and after very rare events for which the capability and availability of design features installed on the site might be affected[22]. The aim of the use of non-permanent equipment is to restore safety functions that have been lost, but its use should not be the regular means for coping in the short term phase of design basis accidents or for design extension conditions (see also paras 7.51 and 7.64 of SSG-2 (Rev. 1) [9]).

5.6. To meet the requirements established in SSR-2/1 (Rev. 1) [1] (see also paras 5.2 and 5.3), levels of hazards exceeding those considered for design (i.e. those derived from the hazard evaluation for the site) should be considered and their consequences should be evaluated as part of the defence in depth approach. For natural external hazards, it is not always possible to have sufficient confidence in the frequency of occurrence of a certain level of hazard for the definition of a design basis level. In this case, rather than trying to associate levels with frequencies, the level of natural hazards exceeding the level considered for design, derived from the hazard evaluation for the site, should be defined by the addition of an adequate margin. The behaviour of structures, systems and

---

[21] These requirements in SSR-2/1 (Rev. 1) [1] were the result of feedback from the accident at the Fukushima Daiichi nuclear power plant. Therefore, these measures were primarily introduced with the occurrence of extreme external hazards in mind.

[22] Further considerations related to non-permanent equipment are provided in SSG-54 [15].

components due to loading parameters resulting from these levels should be assessed with regard to potential use of non-permanent equipment (e.g. coping time for deployment).

5.7. An evaluation should be conducted to demonstrate that the plant would be able to cope with an external hazard of a severity exceeding the levels considered for design, derived from the hazard evaluation for the site, on the basis of both of the following:

(a) An analysis of adequate design margins of the structures, systems and components that are necessary to reach a safe state, against the resulting higher loads that might be present;
(b) An analysis of the use of non-permanent equipment to restore the necessary safety functions after the main effects of the hazard have passed.

5.8. For each relevant scenario involving an external hazard of a level exceeding the level considered for design, derived from the hazard evaluation for the site, the evaluation should identify limitations on the response capabilities of the plant and a strategy should be defined to cope with these limitations. The evaluation should also identify the various coping provisions, accident management measures and equipment (i.e. fixed or non-permanent equipment stored on the site or off the site) that will be used to restore the safety functions and to reach and maintain a safe state. The evaluation should include the following:

(a) A robustness analysis of a relevant set of items important to safety to estimate the extent to which those items would be able to withstand levels of hazards exceeding those considered for design;
(b) An assessment of the extent to which the nuclear power plant would be able to withstand a loss of the safety functions without there being unacceptable radiological consequences for the public and the environment;
(c) The coping strategies to limit and mitigate the consequences of scenarios that could lead to a loss of relevant safety functions;
(d) An estimate of the necessary resources (i.e. human resources, equipment, logistics and communication) to confirm the feasibility of the coping strategies;
(e) A demonstration that the time available before a safety function is lost provides a sufficient margin over the time needed to perform all necessary actions to restore the safety function.

5.9. Some aspects of the use of non-permanent equipment and the associated safety assessment cannot be fully considered in detail at the design stage and should be considered in the commissioning and operation stages. However, specific provisions for the use of non-permanent equipment to ensure the radiation protection of operating personnel should be considered at the design stage of new nuclear power plants or during the implementation of modifications, where applicable, for nuclear power plants designed to earlier standards.

5.10. The evaluation should consider the possibility that multiple units at the same site could be simultaneously affected by a level of external hazards exceeding those considered for design, derived from the hazard evaluation for the site, including natural external hazards such as earthquakes. This evaluation should be used to define the amount of non-permanent equipment needed.

5.11. The plant response and the coping strategies in relation to the deployment, installation and use of non-permanent equipment for natural external hazards exceeding the levels considered for design should be assessed on the basis of a realistic approach and should be supplemented where relevant (e.g. in the case of cliff edge effects) by sensitivity analyses where assumptions in the modelling or where important operator actions are identified as essential factors for the credibility of the strategy.

5.12. The coping strategies should be defined, and the associated coping provisions in relation to the deployment, installation and use of non-permanent equipment should be specified and designed taking into account the possible scenarios, in accordance with para. 5.8.

5.13. To make the coping strategies more reliable, an adequate balance between fixed equipment and non-permanent equipment should be implemented. This balance should be defined in accordance with the period of time for which each coping strategy will need to be implemented (the 'coping time'), the time needed for the installation of the non-permanent equipment, the flexibility of using equipment for different purposes, human reliability, the availability of human resources and the total number of operator actions needed for the whole coping strategy. The use of fixed equipment should be preferred for the implementation of short term actions.

5.14. The use of non-permanent equipment should be such that the time needed for the installation and putting into service of the equipment is less than the defined coping time, with a specified margin allowed for time sensitive operator actions. Appropriate time margins should be established for implementing operator actions before the occurrence of a cliff edge effect. This time period should be derived, where possible, on the basis of times recorded during drills or other approaches for validating operator actions. The ability to deliver and operate non-permanent equipment on time under adverse conditions at the site should also be demonstrated, particularly for events that could involve significant degradation of infrastructure and roads caused by extreme hazards on the site and off the site. Consideration should be given to storing non-permanent equipment at a distance from the units in case of some extreme hazards.

5.15. The installation and use of non-permanent equipment should be documented, and comprehensive training, testing and drills should be conducted periodically to maintain operator proficiency in the use of the equipment and associated procedures. To the extent practicable, drills should consider the conditions of real emergencies.

5.16. Once the coping strategies have been defined and validated, guidance for operators, as well as the technical basis of the strategies, should be established and documented (e.g. in emergency operating procedures or severe accident management guidelines).

5.17. To ensure the success and reliability of the coping strategies, the performance criteria of the necessary coping provisions should be specified, and equipment should be designed and, when relevant, qualified in accordance with appropriate standards to ensure its functionality during and after conditions caused by an extreme external hazard or other extreme conditions.

5.18. The appropriateness of the coping strategies and coping provisions, the feasibility of their implementation under environmental conditions caused by external hazards exceeding the levels considered for design, and the radiological consequences of the accident should all be evaluated.

REFERENCES

[1]   INTERNATIONAL ATOMIC ENERGY AGENCY, Safety of Nuclear Power Plants: Design, IAEA Safety Standards Series No. SSR-2/1 (Rev. 1), IAEA, Vienna (2016).
[2]   INTERNATIONAL ATOMIC ENERGY AGENCY, IAEA Nuclear Safety and Security Glossary: Terminology Used in Nuclear Safety, Nuclear Security, Radiation Protection and Emergency Preparedness and Response, 2022 (Interim) Edition, IAEA, Vienna (2022), https://doi.org/10.61092/iaea.rrxi-t56z
[3]   INTERNATIONAL ATOMIC ENERGY AGENCY, Safety Assessment for Facilities and Activities, IAEA Safety Standards Series No. GSR Part 4 (Rev. 1), IAEA, Vienna (2016).

[4] EUROPEAN ATOMIC ENERGY COMMUNITY, FOOD AND AGRICULTURE ORGANIZATION OF THE UNITED NATIONS, INTERNATIONAL ATOMIC ENERGY AGENCY, INTERNATIONAL LABOUR ORGANIZATION, INTERNATIONAL MARITIME ORGANIZATION, OECD NUCLEAR ENERGY AGENCY, PAN AMERICAN HEALTH ORGANIZATION, UNITED NATIONS ENVIRONMENT PROGRAMME, WORLD HEALTH ORGANIZATION, Fundamental Safety Principles, IAEA Safety Standards Series No. SF-1, IAEA, Vienna (2006), https://doi.org/10.61092/iaea.hmxn-vw0a

[5] INTERNATIONAL ATOMIC ENERGY AGENCY, Design of the Reactor Coolant System and Associated Systems for Nuclear Power Plants, IAEA Safety Standards Series No. SSG-56, IAEA, Vienna (2020).

[6] INTERNATIONAL ATOMIC ENERGY AGENCY, Design of the Reactor Containment and Associated Systems for Nuclear Power Plants, IAEA Safety Standards Series No. SSG-53, IAEA, Vienna (2019).

[7] INTERNATIONAL ATOMIC ENERGY AGENCY, Design of Electrical Power Systems for Nuclear Power Plants, IAEA Safety Standards Series No. SSG-34, IAEA, Vienna (2016).

[8] INTERNATIONAL ATOMIC ENERGY AGENCY, Design of Instrumentation and Control Systems for Nuclear Power Plants, IAEA Safety Standards Series No. SSG-39, IAEA, Vienna (2016).

[9] INTERNATIONAL ATOMIC ENERGY AGENCY, Deterministic Safety Analysis for Nuclear Power Plants, IAEA Safety Standards Series No. SSG-2 (Rev. 1), IAEA, Vienna (2019).

[10] INTERNATIONAL ATOMIC ENERGY AGENCY, Development and Application of Level 1 Probabilistic Safety Assessment for Nuclear Power Plants, IAEA Safety Standards Series No. SSG-3 (Rev. 1), IAEA, Vienna (in press).

[11] INTERNATIONAL ATOMIC ENERGY AGENCY, Development and Application of Level 2 Probabilistic Safety Assessment for Nuclear Power Plants, IAEA Safety Standards Series No. SSG-4, IAEA, Vienna (2010). (A revision of this publication is in preparation.)

[12] FOOD AND AGRICULTURE ORGANIZATION OF THE UNITED NATIONS, INTERNATIONAL ATOMIC ENERGY AGENCY, INTERNATIONAL CIVIL AVIATION ORGANIZATION, INTERNATIONAL LABOUR ORGANIZATION, INTERNATIONAL MARITIME ORGANIZATION, INTERPOL, OECD NUCLEAR ENERGY AGENCY, PAN AMERICAN HEALTH ORGANIZATION, PREPARATORY COMMISSION FOR THE COMPREHENSIVE NUCLEAR-TEST-BAN TREATY ORGANIZATION, UNITED NATIONS ENVIRONMENT PROGRAMME, UNITED NATIONS OFFICE FOR THE COORDINATION OF HUMANITARIAN AFFAIRS, WORLD HEALTH ORGANIZATION, WORLD METEOROLOGICAL ORGANIZATION, Preparedness and Response for a Nuclear or Radiological Emergency, IAEA Safety Standards Series No. GSR Part 7, IAEA, Vienna (2015), https://doi.org/10.61092/iaea.3dbe-055p

[13] INTERNATIONAL ATOMIC ENERGY AGENCY, Radiation Protection Aspects of Design for Nuclear Power Plants, IAEA Safety Standards Series No. SSG-90, IAEA, Vienna (in press).

[14] INTERNATIONAL ATOMIC ENERGY AGENCY, UNITED NATIONS ENVIRONMENT PROGRAMME, Radiation Protection of the Public and the Environment, IAEA Safety Standards Series No. GSG-8, IAEA, Vienna (2018).

[15] INTERNATIONAL ATOMIC ENERGY AGENCY, Accident Management Programmes for Nuclear Power Plants, IAEA Safety Standards Series No. SSG-54, IAEA, Vienna (2019).

[16] INTERNATIONAL ATOMIC ENERGY AGENCY, Site Evaluation for Nuclear Installations, IAEA Safety Standards Series No. SSR-1, IAEA, Vienna (2019).

Annex I

EXAMPLES OF CASES OF PRACTICAL ELIMINATION

I–1. This annex illustrates potential examples of cases of practical elimination. It needs to be noted that both the list of examples as well as the associated content differ among Member States.

FAILURE OF A LARGE COMPONENT IN THE REACTOR COOLANT SYSTEM

I–2. A sudden mechanical failure of a single large component in the reactor coolant system could initiate an event in which reactor cooling would be lost in a short time and a pressure wave or a missile would damage the containment boundary. The safety provisions for defence in depth would not be effective in such a situation and an early radioactive release or a large radioactive release could follow. This is a very exceptional type of initiating event that safety systems and safety features are not designed to mitigate and therefore it needs to be demonstrated with high confidence that the likelihood of such an initiating event occurring would be so low that it can be excluded (i.e. practically eliminated) from consideration. This is particularly important for the reactor vessel, in which a break would eliminate the capability of holding and cooling the core. In addition, the likelihood of a failure of the pressurizer or the steam generator shell needs to be shown to be extremely low, or alternatively it needs to be demonstrated that a failure of the pressurizer or the steam generator shell would not lead to unacceptable consequences for the containment.

I–3. The safety demonstration needs to be especially robust and the corresponding assessment suitably demanding, so that an engineering judgement can be made for the following key aspects of each large component in the reactor coolant system:

(a) The most suitable composition of materials needs to be selected.
(b) The metal component or structure needs to be as defect-free as possible.
(c) The metal component or structure needs to be tolerant of defects.
(d) The mechanisms of growth of defects need to be known.
(e) Design provisions and suitable operating practices need to be in place to minimize thermal fatigue, stress corrosion, embrittlement, pressurized thermal shock and overpressurization of the primary circuit.

(f)   Continuous leak detection capability is needed during pressurized operation.
(g)   Effective in-service inspection and surveillance and chemistry control programmes need to be in place during the manufacture, construction, commissioning and operation of the equipment, to detect any defects or degradation mechanisms and to ensure that equipment properties are preserved over the lifetime of the plant.

I–4.   In addition, evidence needs to be provided to demonstrate that the necessary integrity of large components of the reactor coolant system will be maintained for the most demanding situations.

I–5.   Several sets of well established technical standards are available for ensuring the reliability of large pressure vessels, and the demonstration of practical elimination of failures of the pressure vessel has to be based on the rigorous application of these technical standards. Such technical standards also provide instructions for the verification of the state of the pressure vessel during the lifetime of the vessel.

I–6.   The practical elimination of failures of large components is thus achieved by the first level of defence in depth and does not rely on the subsequent levels of defence in depth.

I–7.   The demonstration, with a high level of confidence, of a low likelihood of failure could be supplemented by a probabilistic fracture mechanics assessment, which is a widely recognized and commonly used technique. Probabilistic assessment in the demonstration of practical elimination, especially in this case, is not to be restricted to the use of Boolean reliability models (e.g. fault trees, event trees) or failure rates derived from the statistical analysis of observed catastrophic failures. Probabilistic fracture mechanics assessments address aspects such as material fracture toughness and weld residual stress, which in turn consider deterministic analysis, engineering judgement and the measurements of monitored values.

FAST REACTIVITY INSERTION ACCIDENT IN A LIGHT WATER REACTOR

I–8.   Fast reactivity accidents can be very energetic and have a potential to destroy the fuel, fuel cladding and other barriers. As far as practicable, the prevention of such accidents is to be ensured at the first level of defence in depth by proper design of the reactor coolant system and the core, or at the third level of defence in depth by provision of two diverse, independent means of shutdown.

I–9. The first level of defence in depth may be provided by the nuclear characteristics of the reactor core (such as the negative reactivity coefficient in light water reactors), which, under all possible combinations of reactor power, neutron absorber concentration, coolant pressure and temperature, suppress any increase in reactor power during any disturbances and eliminate any uncontrolled reactivity excursion. Therefore, this is a case of demonstration of practical elimination by physical impossibility of the event sequence.

I–10. An uncontrolled reactivity excursion could potentially be caused by the sudden insertion of a cold or underborated water slug into a reactor core. By design, the accident could be considered as eliminated by demonstrating that only a limited volume of unborated water could be injected, which does not allow this effect to happen. The accident could also be considered as eliminated by demonstrating that sufficient negative reactivity coefficient exists for possible combinations of the reactor power and coolant pressure and temperature for the core cycle. Nevertheless, all potential risks of sudden changes in the coolant properties need to be identified and prevented by design provisions. In this case, the demonstration of practical elimination is because the event sequence is considered physically impossible to occur.

I–11. Therefore, the demonstration of practical elimination relies primarily on the impossibility of reactivity excursions through a core design with overall small or negative reactivity coefficients, supported by other design measures to avoid or limit excursions of reactivity, which can be evaluated deterministically and probabilistically as appropriate to demonstrate that the conditions are extremely unlikely to occur.

I–12. A more complex situation could arise, however, if criticality can be reached during a severe accident. This has been a topic of concern for specific core meltdown scenarios in reactors, for which the control rod material has a lower melting point and eutectic formation temperature than the fuel rods. A potentially hazardous scenario might occur if the reactor vessel were reflooded with unborated water in a situation when the control rods have relocated downwards but the fuel rods are still in their original position. This could result in recriticality of the fuel, likely resulting in the generation of additional heat on a continuing or intermediate basis, depending on the presence of water. This is again an aspect to be analysed by considering the design provisions and severe accident management features together, in order to be able to demonstrate that the plant sequence has been practically eliminated because it is considered, with a high level of confidence, to be extremely unlikely to occur.

DIRECT CONTAINMENT HEATING

I–13. In a pressure vessel reactor, core meltdown at high pressure could cause a violent discharge of molten corium material into the containment atmosphere and this would result in direct containment heating from the hot melt and exothermic chemical reactions. Plant event sequences involving high pressure core melt therefore need to be practically eliminated by design provisions to depressurize the reactor coolant system when a meltdown is found unavoidable, so that the conditions are considered, with a high level of confidence, to be extremely unlikely to occur.

I–14. In a pressurized heavy water reactor, by contrast, direct containment heating due to ejection of the molten corium at high pressure is practically eliminated because pressure tubes would fail rapidly at high fuel temperature. This would depressurize the primary system before significant core melting can occur. This is a case of practical elimination of the event sequence owing to its physical impossibility.

I–15. Any high pressure core meltdown scenario would evidently be initiated by a small coolant leak or boiling of the coolant and release of steam through a safety or relief valve. For such situations, design provisions need to be in place to ensure, with a high level of confidence, that such small coolant leaks or boiling of the coolant would instead result, with a high reliability, in a low pressure core melt sequence, so that high pressure core melt conditions can be practically eliminated. The depressurization needs to be such that very low pressure can be achieved before any discharge of molten corium from the reactor vessel can take place. In addition, it is important that dynamic loads from depressurization do not pose a risk to the integrity of the containment. Design provisions need to be in place to ensure, with a high level of confidence, that any high pressure core meltdown scenario does not occur.

I–16. Dedicated depressurization systems have been installed in existing plants and designed for new plants. In pressurized water reactors, such systems are based on simple and robust devices and straightforward actions by operating personnel that eliminate the risk of erroneous automatic depressurization but provide adequate time to act if the need arises. In boiling water reactors, the existing steam relief systems generally provide means for depressurization, with possibly some modifications in valve controls to also ensure reliable valve opening and open valve positions at very low pressures.

I–17. A deterministic safety analysis is necessary to demonstrate the effectiveness of the depressurization system in preventing direct containment heating. Traditional probabilistic safety assessment techniques are adequate to demonstrate a high reliability of the depressurization systems, including the initiation of the systems by operating personnel. In this way, direct containment heating could be demonstrated, with a high level of confidence, to be extremely unlikely to occur, based on a combined deterministic and probabilistic assessment of specific design provisions.

LARGE STEAM EXPLOSION

I–18. The interaction of the reactor core melt with water, known as fuel–coolant interaction, is a complex technical issue involving a number of thermohydraulic and chemical phenomena. Fuel–coolant interactions might occur in-vessel, during flooding of a degraded core or if a molten core relocates into the lower head filled with water. Such interactions might also occur ex-vessel, if molten core debris is ejected into a flooded reactor cavity after vessel failure. Each of the scenarios might lead to an energetic fuel–coolant interaction, commonly known as 'steam explosion', which represents a potentially serious challenge to the integrity of the reactor vessel and/or the containment.

I–19. The conditions that trigger a steam explosion and the energy of explosion in various situations have been widely studied in reactor safety research programmes. The risks of steam explosion cannot be fully eliminated for all core meltdown scenarios in which molten core might drop into water.

I–20. For the practical elimination of steam explosions that could damage the integrity of the containment, the preferred method is to avoid the dropping of molten core into water for all conceivable accident scenarios. Such an approach is used in some pressurized water reactors where the reliability of external cooling of the molten core has been proven and in some new reactors with a separate core catcher. In some existing boiling water reactors and in some new designs of boiling water reactors, the molten core would drop into a pool below the reactor vessel in all severe accident scenarios and would be solidified and cooled in the pool. In all such circumstances in which the molten core drops into water, it needs to be proven with arguments based on the physical phenomena involved in the respective scenarios that the risk of steam explosion damaging the integrity of the containment has been practically eliminated owing to the physical impossibility of the event sequence.

EXPLOSION OF COMBUSTIBLE GASES: HYDROGEN AND CARBON MONOXIDE

I–21. Hydrogen combustion is a very energetic phenomenon, and a fast combustion reaction (detonation) involving a sufficient amount of hydrogen would cause a significant threat to the integrity of the containment. Dedicated means to prevent the generation of hydrogen and its accumulation at critical concentrations, and to eliminate hydrogen detonation, are needed at all nuclear power plants, although different means are preferred for different plant designs.

I–22. In boiling water reactor containments, which are all relatively small, the main means of protection against hydrogen generation and accumulation is the filling of the containment with inert nitrogen gas during power operation. In large, pressurized water reactor containments, the current practice is to use passive catalytic recombiners or other devices that control the rate of the oxygen and hydrogen recombination against hydrogen detonation.

I–23. It is also necessary to ensure and confirm with analysis and tests that the circulation of gases and steam inside the containment provides proper conditions for hydrogen recombination and eliminates excessive local hydrogen concentration, taking into account that the risk of hydrogen detonation increases if steam providing inertization is condensed.

I–24. The consequences of hydrogen combustion will depend on the highest conceivable rate and the total amount of hydrogen generation inside the containment. Some core catchers that are currently installed in nuclear power plants can significantly reduce or even eliminate ex-vessel hydrogen generation in an accident when the molten core has dropped into the catcher, and this could also considerably reduce the total amount of hydrogen generated inside the containment.

I–25. In particular, the design provisions for preventing hydrogen detonation need to be assessed in order to demonstrate the practical elimination of this phenomenon. This assessment also includes the consideration of (i) the appropriate selection of materials allowing a limited amount of hydrogen generation during a severe accident and (ii) the hydrogen propagation and mixing inside the containment.

I–26. Carbon monoxide can be generated in a severe accident if molten core discharged from the reactor vessel interacts with concrete structures. The amount and the timing of carbon monoxide generated depend on the particular core melt scenario, the type of concrete and geometric factors. Mixtures of carbon monoxide and air can also be explosive, although this chemical reaction is less energetic than hydrogen combustion and the burning velocity is also lower. Therefore, the contribution of carbon monoxide to the risks to the integrity of the containment has generally received less attention. However, the presence of carbon monoxide increases the combustible gas inventory in the containment and also influences flammability limits and burning velocities of hydrogen. Therefore, the influence of carbon monoxide needs to be considered so as to demonstrate the practical elimination of hydrogen combustion. A design provision to minimize the impact of carbon monoxide is the use of concrete with low limestone content.

LONG TERM LOSS OF CONTAINMENT HEAT REMOVAL

I–27. In a situation where core decay heat cannot be removed by heat transfer systems to outside the containment and removed further to an ultimate heat sink, or in a severe accident where the core is molten and is generating steam inside the containment, cooling of the containment atmosphere is a preferred means for preventing its overpressure.

I–28. There are several examples, both from existing plants and from new plant designs, of dedicated robust containment cooling systems that are independent of safety systems and might be capable of supporting the demonstration of practical elimination of containment rupture by overpressure.

I–29. An alternative to the cooling of the containment is the elimination of containment overpressure by means of venting. This is necessary especially in some boiling water reactors where the size of the containment is small and pressure limitation might be needed for design basis accidents and design extension conditions with core melt. The venting systems in existing plants prevent overpressurization at the cost of some radioactive release involved in the venting, also in the case that the venting is filtered. However, these might be acceptable strategies for severe accident management if technically justified given the risk levels and an appropriate assessment of the decontamination factors for the strategy.

I–30. Containment venting avoids a risk to the integrity of the containment resulting from overpressurization, but stabilization of the core and the cooling of the containment are still necessary in the longer term.

I–31. The safety demonstration needs to be based on the capability and reliability of the specific measures implemented in the design to cope with the severe accident phenomena. Level 2 probabilistic safety assessment can be used to demonstrate the very low probability of plant event sequences that could lead to a large radioactive release (i.e. the practical elimination of long term loss of containment heat removal as it is considered, with a high level of confidence, to be extremely unlikely to arise).

CONTAINMENT PENETRATION BY INTERACTION WITH THE MOLTEN CORE

I–32. In a severe accident in which the core has melted through the reactor vessel, it is possible that the integrity of the containment could be breached if the molten core is not sufficiently cooled. In addition, interactions between the core debris and concrete can generate large quantities of additional combustible gases, hydrogen and carbon monoxide, as well as other non-condensable gases, which could also contribute to eventual overpressure failure of the containment.

I–33. Alternative means have been developed and verified in extensive severe accident research programmes in this area conducted in several States and also with international cooperation. The suggested means include the following:

(a) Keeping the molten core inside the reactor vessel by cooling the vessel from outside;
(b) Installing a dedicated system or device that would catch and cool the molten core as soon as it has penetrated the reactor vessel wall.

I–34. In both approaches, cooling of the molten core generates steam inside the containment, and it is also necessary to provide features for heat removal from the containment that are independent, to the extent practicable, of those used in more frequent accidents.

I–35. While probabilistic safety assessment can play a role in assessing the reliability of establishing external reactor vessel cooling or the core catcher cooling (if provided), the demonstration of the practical elimination of melt through the containment boundary relies extensively on deterministic analysis of the design provisions, to demonstrate that such containment penetration can be considered, with a high level of certainty, to be extremely unlikely to arise.

SEVERE ACCIDENTS WITH CONTAINMENT BYPASS

I–36. Containment bypass can occur in different ways, such as through circuits connected to the reactor coolant system that exit the containment or as a result of defective steam generator tubes (for pressurized water reactors). Severe accident sequences with non-isolated penetrations connecting the containment atmosphere to the outside and severe accident sequences during plant shutdown with the containment open also need to be considered as containment bypass scenarios. Failures of lines exiting the containment and connected to the primary system, including steam generator tube ruptures, are at the same time accident initiators, whereas other open penetrations only constitute a release path in accident conditions. Nevertheless, all these plant event sequences have to be practically eliminated by design provisions such as adequate piping design pressure and isolation mechanisms.

I–37. The safety demonstration for elimination of bypass sequences includes a systematic review of all potential containment bypass sequences and covers all containment penetrations.

I–38. Requirement 56 of IAEA Safety Standards Series No. SSR-2/1 (Rev. 1), Safety of Nuclear Power Plants: Design [I–1], establishes the minimum isolation requirements for various kinds of containment penetration. The requirement addresses aspects of leaktightness and leak detection, redundancy, and automatic actuations, as appropriate. Specific provisions are given also for interfacing failures in the reactor coolant system. National regulations address in more detail the applicable provisions for containment isolations and prevention of containment bypass or loss of cooling accidents in interface systems.

I–39. Based on the implementation of the design requirements or specific national regulations and the in-service inspection and surveillance practices at the plant, the analysis has to assess the frequency of bypassing mechanisms. This analysis, although of probabilistic nature, needs to combine aspects of engineering judgement and deterministic analysis in the probabilistic calculations, and always to be based on the redundancy and robustness of the design, the application of relevant design rules (e.g. fail-safe actuation), as well as the pertinent inspection provisions and operational practices, as was done in previous cases. While the analysis of isolation of containment penetrations or steam generators is amenable to conventional fault tree and event tree analyses, with due consideration of failures in power supplies, isolation signals and operator actions, other analysis aspects might involve the use of other probabilistic methods together with deterministic methods and engineering judgement to demonstrate the practical elimination of containment bypass. This would lead to a defensible low frequency estimate of the bypass mechanisms associated with each penetration. In addition, the reliability of design provisions for the isolation of bypass paths based on conventional probabilistic assessments would complement the demonstration that plant event sequences with containment bypass have been practically eliminated.

SIGNIFICANT FUEL DEGRADATION IN THE SPENT FUEL POOL

I–40. Facilities for spent fuel storage need to be designed to ensure that plant event sequences that could lead to an early radioactive release or a large radioactive release to the environment are practically eliminated. To this end, it is necessary to ensure that spent fuel stored in a pool is always kept covered by an adequate layer of water. This involves the following:

(a) Providing a pool structure that is designed to protect against all conceivable internal hazards and external hazards that could damage its integrity.
(b) Avoiding the siphoning of water out of the pool.
(c) Providing sufficiently reliable means for pool cooling that eliminate the possibility of a long lasting loss of cooling function (i.e. for the time needed to boil off the water). An example is the application of redundancy, diversity and independence (see para. 3.7 of IAEA Safety Standards Series No. SSG-63, Design of Fuel Handling and Storage Systems for Nuclear Power Plants [I–2]).
(d) Providing reliable instrumentation for pool level monitoring.
(e) Providing appropriate reliable means to compensate for any losses of water inventory.

I–41. The risks of mechanical fuel failures need to be eliminated by the following means:

(a) A design that ensures that movements of heavy lifts (e.g. transport casks) above the spent fuel stored in the pool are avoided;
(b) Structures that eliminate the possibility of heavy lifts dropping on top of the fuel.

I–42. In designs where the spent fuel pool is outside the containment, the uncovering of the fuel would lead to fuel damage and a large radioactive release could not be prevented. Means to evacuate the hydrogen would prevent explosions that could cause further damage and prevent a later reflooding and cooling of the fuel. Therefore, it is necessary to ensure through design provisions that the uncovering of spent fuel elements has been practically eliminated.

I–43. In some designs, the spent fuel pool is located inside the containment. In this case, even though spent fuel damage would not lead directly to a large radioactive release, the amount of hydrogen generated by a large number of fuel elements and the easy penetration of the pool liner by the molten fuel without means to stabilize it, among other harsh effects, could eventually lead to a large radioactive release. Therefore, it is also necessary to ensure through design provisions that, in this case also, the uncovering of spent fuel elements has been practically eliminated.

REFERENCES TO ANNEX I

[I–1] INTERNATIONAL ATOMIC ENERGY AGENCY, Safety of Nuclear Power Plants: Design, IAEA Safety Standards Series No. SSR-2/1 (Rev. 1), IAEA, Vienna (2016).
[I–2] INTERNATIONAL ATOMIC ENERGY AGENCY, Design of Fuel Handling and Storage Systems for Nuclear Power Plants, IAEA Safety Standards Series No. SSG-63, IAEA, Vienna (2020).

Annex II

APPLICATION OF THE CONCEPTS OF DESIGN EXTENSION CONDITIONS AND PRACTICAL ELIMINATION TO NUCLEAR POWER PLANTS DESIGNED TO EARLIER STANDARDS

II–1. Paragraph 1.3 of IAEA Safety Standards Series No. SSR-2/1 (Rev. 1), Safety of Nuclear Power Plants: Design [II–1], states:

"It might not be practicable to apply all the requirements of this Safety Requirements publication to nuclear power plants that are already in operation or under construction. In addition, it might not be feasible to modify designs that have already been approved by regulatory bodies. For the safety analysis of such designs, it is expected that a comparison will be made with the current standards, for example as part of the periodic safety review for the plant, to determine whether the safe operation of the plant could be further enhanced by means of reasonably practicable safety improvements."

This implies that (i) the capability of existing plants to accommodate accident conditions not considered in their current design basis and (ii) the practical elimination of plant event sequences that could lead to an early radioactive release or a large radioactive release need to be assessed as part of the periodic safety review processes, with the objective of further improving the level of safety, where reasonably practicable.

II–2. The concepts of design extension conditions and practical elimination of plant event sequences that could lead to an early radioactive release or a large radioactive release are not new. In fact, the concept of practical elimination was already introduced in the 2004 IAEA Safety Guide for the design of the reactor containment[1], and both concepts might have been applied partially in the design of some existing nuclear power plants, although not necessarily in a systematic way. Over time, design features to cope with conditions such as station blackout or anticipated transients without scram have been introduced in many nuclear

---

[1] See para. 6.5 of INTERNATIONAL ATOMIC ENERGY AGENCY, Design of Reactor Containment Systems for Nuclear Power Plants, IAEA Safety Standards Series No. NS-G-1.10, IAEA, Vienna (2004), which has been superseded by INTERNATIONAL ATOMIC ENERGY AGENCY, Design of the Reactor Containment and Associated Systems for Nuclear Power Plants, IAEA Safety Standards Series No. SSG-53, IAEA, Vienna (2019) [II–2].

power plants. Some event sequences that could lead to an early radioactive release or a large radioactive release have also been addressed in many designs already, although a specific demonstration of the practical elimination of such plant event sequences has not been carried out.

II–3. In relation to practical elimination, a number of measures might have been taken, for example, for the prevention of a break in the reactor pressure vessel, for fast reactivity insertion accidents or for severe fuel degradation in the spent fuel pool. However, a demonstration that the existing safety provisions are sufficient to claim the practical elimination of such plant event sequences might not have been conducted in the way required by SSR-2/1 (Rev. 1) [II–1] and as recommended in this Safety Guide.

II–4. However, an accident condition commonly considered as a design extension condition in a new nuclear power plant (e.g. station blackout, anticipated transient without scram) can only be considered a design extension condition for an existing nuclear power plant if safety features have been introduced in the original design of the existing plant to mitigate the consequences of this condition. For the case of station blackout, an alternate power source capable of supplying power in due time to essential loads over a sufficient time period until external or emergency power is recovered would be an example of an original design safety feature. Likewise, for anticipated transient without scram, additional design features capable of rendering the reactor subcritical in the case of failure in the insertion of control rods would need to be included in the original design. Without such additional design features in the original design, these accident conditions would need to be considered to be beyond the design basis of the plant.

II–5. Generally, it is expected that during a periodic safety review or a reassessment of plant safety, or as part of a request for lifetime extension or similar processes, the feasibility of reasonable safety improvements in relation to design extension conditions and practical elimination would be considered. There can, however, be constraints on installing the same type of design features as commonly implemented in the design of new nuclear power plants, especially for design extension conditions with core melting such as the implementation of the ex-vessel melt retention or in-vessel corium cooling strategies in pressurized water reactor designs. In the same context, there can be constraints on ensuring the independence of safety provisions relating to the different levels of defence in depth.

II–6. Safety provisions for design extension conditions and also design features for the practical elimination of plant event sequences that could lead to an early radioactive release or a large radioactive release are addressed in several Safety Guides related to the design of plant systems, including SSG-53 [II–2] and IAEA Safety Standards Series Nos SSG-56, Design of the Reactor Coolant System and Associated Systems for Nuclear Power Plants [II–3]; SSG-34, Design of Electrical Power Systems for Nuclear Power Plants [II–4]; and SSG-39, Design of Instrumentation and Control Systems for Nuclear Power Plants [II–5]. SSG-53 [II–2] encompasses most of the design features for design extension conditions with core melting and addresses the plant event sequences to be considered for practical elimination. SSG-53 [II–2] also contains an appendix in relation to nuclear power plants designed to earlier standards that provides recommendations for the upgrading of the plant design in relation to these aspects.

II–7. Safety systems of existing plants were designed for design basis accidents, without account being taken in the design of the prevention and mitigation of more severe accidents. However, the conservative deterministic approaches originally followed in the design might have resulted in the capability to withstand some situations more severe than those originally included in the design basis for existing plants. As indicated in para. 3.22 of this Safety Guide on design extension conditions without significant fuel degradation, for postulated initiating events less frequent than those considered for design basis accidents, it can be acceptable to demonstrate that some safety systems would be capable of and qualified for mitigating the consequences of such events if best estimate analyses and less conservative assumptions are used. For existing nuclear power plants, this is a possibility to demonstrate the capability for mitigation of design extension conditions not originally postulated in the design, such as a multiple rupture of steam generator tubes. Existing nuclear power plants could also extend the capability of safety systems to be capable of mitigation of some design extension conditions, in accordance with para. 5.27 of SSR-2/1 (Rev. 1) [II–1].

II–8. The consideration of external events of a magnitude exceeding the original design basis derived from the hazard evaluation for the site, as addressed in Section 5, is to be considered. While for new nuclear power plants the mitigation of design extension conditions is generally expected to be accomplished by permanent design features, and the use of non-permanent equipment is intended only for very unlikely external events of a magnitude exceeding the original design basis, for existing nuclear power plants the use of non-permanent equipment with adequate connection features can be the only reasonable improvement in some cases. Relying on non-permanent equipment might be adequate provided there is a justification to demonstrate that the coping time to prevent the loss of the safety function that the equipment is intended to fulfil is long enough to connect and put into service the equipment under the conditions associated with the accident. The recommendations in this regard provided in Section 5 are relevant. Non-permanent equipment that would be necessary to reduce further the consequences of events that cannot be mitigated by the installed plant capabilities needs to be stored and protected to ensure its availability when necessary, with account taken of possible restricted access owing to external events (e.g. flooding, damaged roads), and its operability needs to be verified.

REFERENCES TO ANNEX II

[II–1] INTERNATIONAL ATOMIC ENERGY AGENCY, Safety of Nuclear Power Plants: Design, IAEA Safety Standards Series No. SSR-2/1 (Rev. 1), IAEA, Vienna (2016).

[II–2] INTERNATIONAL ATOMIC ENERGY AGENCY, Design of the Reactor Containment and Associated Systems for Nuclear Power Plants, IAEA Safety Standards Series No. SSG-53, IAEA, Vienna (2019).

[II–3] INTERNATIONAL ATOMIC ENERGY AGENCY, Design of the Reactor Coolant System and Associated Systems for Nuclear Power Plants, IAEA Safety Standards Series No. SSG-56, IAEA, Vienna (2020).

[II–4] INTERNATIONAL ATOMIC ENERGY AGENCY, Design of Electrical Power Systems for Nuclear Power Plants, IAEA Safety Standards Series No. SSG-34, IAEA, Vienna (2016).

[II–5] INTERNATIONAL ATOMIC ENERGY AGENCY, Design of Instrumentation and Control Systems for Nuclear Power Plants, IAEA Safety Standards Series No. SSG-39, IAEA, Vienna (2016).

DEFINITION

Practical elimination

The concept of practical elimination applies to plant event sequences that could lead to unacceptable consequences (i.e. an early radioactive release or a large radioactive release) that cannot be mitigated by reasonably practicable means. Practical elimination implies that those plant event sequences have to be demonstrated to be either physically impossible or, with a high level of confidence, extremely unlikely to arise by implementing safety provisions in the form of design and operational features.

Note: Practical elimination is part of a general approach to design safety and complements the adequate implementation of the concept of defence in depth.

CONTRIBUTORS TO DRAFTING AND REVIEW

Bernard, M.	Électricité de France, France
Buttery, N.	European Nuclear Installations Safety Standards
Courtin, E.	World Nuclear Association
Dakin, R.	Office for Nuclear Regulation, United Kingdom
Delfini, G.	Authority for Nuclear Safety and Radiation Protection, Netherlands
Ermolaev, A.	All-Russia Scientific Research Institute for Nuclear Power Plant Operation, Russian Federation
Exley, R.	Office for Nuclear Regulation, United Kingdom
Franovich, M.	Nuclear Regulatory Commission, United States of America
Garis, N.	Swedish Radiation Safety Authority, Sweden
Gyepi-Garbrah, S.	Canadian Nuclear Safety Commission, Canada
Hardwood, C.	Canadian Nuclear Safety Commission, Canada
Ibrahim, M.A.	Nuclear Power Plants Authority, Egypt
Jansen, R.	Authority for Nuclear Safety and Radiation Protection, Netherlands
Järvinen, M.L.	Radiation and Nuclear Safety Authority, Finland
Kim, K.T.	Korea Atomic Energy Research Institute, Republic of Korea
Koski, S.	Teollisuuden Voima Oyj, Finland
Kral, P.	Nuclear Research Institute Řež, Czech Republic
Lignini, F.M.	World Nuclear Association
Luis Hernandez, J.	International Atomic Energy Agency
Müllner, N.A.	Institute of Safety and Risk Sciences, Austria

Nakajima, T.	Nuclear Regulation Authority, Japan
Nünighoff, K.	Gesellschaft für Anlagen- und Reaktorsicherheit gGmbH, Germany
Obenius Mowitz, A.	Swedish Radiation Safety Authority, Sweden
Poulat, B.	Consultant, France
Ranval, W.	European Nuclear Installations Safety Standards
Rodriguez Mate, C.	Nuclear Safety Authority, France
Rogatov, D.	Scientific and Engineering Centre for Nuclear and Radiation Safety, Russian Federation
Schwarz, G.R.	Consultant, Canada
Stoppa, G.	Federal Ministry for the Environment, Nature Conservation, Nuclear Safety and Consumer Protection, Germany
Tas, F.B.	Nuclear Regulatory Authority, Türkiye
Titus, B.A	Nuclear Regulatory Commission, United States of America
Uhrik, P.	Nuclear Regulatory Authority, Slovakia
Virtanen, E.	Radiation and Nuclear Safety Authority, Finland
Wattelle, E.	Institute for Radiation Protection and Nuclear Safety, France
Wong, E.K.Y.	National Environment Agency, Singapore
Yllera, J.	International Atomic Energy Agency

IAEA
International Atomic Energy Agency

ORDERING LOCALLY

IAEA priced publications may be purchased from the sources listed below or from major local booksellers.

Orders for unpriced publications should be made directly to the IAEA. The contact details are given at the end of this list.

NORTH AMERICA

Bernan / Rowman & Littlefield
15250 NBN Way, Blue Ridge Summit, PA 17214, USA
Telephone: +1 800 462 6420 • Fax: +1 800 338 4550
Email: orders@rowman.com • Web site: www.rowman.com/bernan

REST OF WORLD

Please contact your preferred local supplier, or our lead distributor:

Eurospan Group
Gray's Inn House
127 Clerkenwell Road
London EC1R 5DB
United Kingdom

Trade orders and enquiries:
Telephone: +44 (0)176 760 4972 • Fax: +44 (0)176 760 1640
Email: eurospan@turpin-distribution.com

Individual orders:
www.eurospanbookstore.com/iaea

For further information:
Telephone: +44 (0)207 240 0856 • Fax: +44 (0)207 379 0609
Email: info@eurospangroup.com • Web site: www.eurospangroup.com

Orders for both priced and unpriced publications may be addressed directly to:
Marketing and Sales Unit
International Atomic Energy Agency
Vienna International Centre, PO Box 100, 1400 Vienna, Austria
Telephone: +43 1 2600 22529 or 22530 • Fax: +43 1 26007 22529
Email: sales.publications@iaea.org • Web site: www.iaea.org/publications